T0251799

FORAGE CROPS OF THE WORLD

Volume II: Minor Forage Crops

FORAGE CROPS OF THE WORLD

Volume II: Minor Forage Crops

Edited by

Md. Hedayetullah, PhD
Parveen Zaman, PhD

APPLE ACADEMIC PRESS

Apple Academic Press Inc.
3333 Mistwell Crescent
Oakville, ON L6L 0A2
Canada

Apple Academic Press Inc.
9 Spinnaker Way
Waretown, NJ 08758
USA

© 2019 by Apple Academic Press, Inc.

First issued in paperback 2021

Exclusive worldwide distribution by CRC Press, a member of Taylor & Francis Group
No claim to original U.S. Government works

ISBN 13: 978-1-77463-170-6 (pbk)
ISBN 13: 978-1-77188-685-7 (hbk)

ISBN 13: 978-1-77188-686-4 (hbk) (2-volume set)

All rights reserved. No part of this work may be reprinted or reproduced or utilized in any form or by any electric, mechanical or other means, now known or hereafter invented, including photocopying and recording, or in any information storage or retrieval system, without permission in writing from the publisher or its distributor, except in the case of brief excerpts or quotations for use in reviews or critical articles.

This book contains information obtained from authentic and highly regarded sources. Reprinted material is quoted with permission and sources are indicated. Copyright for individual articles remains with the authors as indicated. A wide variety of references are listed. Reasonable efforts have been made to publish reliable data and information, but the authors, editors, and the publisher cannot assume responsibility for the validity of all materials or the consequences of their use. The authors, editors, and the publisher have attempted to trace the copyright holders of all material reproduced in this publication and apologize to copyright holders if permission to publish in this form has not been obtained. If any copyright material has not been acknowledged, please write and let us know so we may rectify in any future reprint.

Trademark Notice: Registered trademark of products or corporate names are used only for explanation and identification without intent to infringe.

Library and Archives Canada Cataloguing in Publication

Forage crops of the world / edited by Md. Hedayetullah, PhD, Parveen Zaman, PhD.

Includes bibliographical references and indexes.
Content: Volume I. Major forage crops -- Volume II. Minor forage crops.
Issued in print and electronic formats.
ISBN 978-1-77188-684-0 (v. 1 : hardcover).--ISBN 978-1-77188-685-7
(v. 2 : hardcover).--ISBN 978-1-77188-686-4 (set : hardcover).--
ISBN 978-1-351-16736-9 (v. 1 : PDF).--ISBN 978-1-351-16728-4
(v. 2 : PDF).--ISBN 978-1-351-16724-6 (set : PDF)

1. Forage plants. I. Hedayetullah, Md., 1982-, editor II. Zaman, Parveen, 1989-, editor

SB193.F67 2018 633.2 C2018-903719-9 C2018-903720-2

CIP data on file with US Library of Congress

Apple Academic Press also publishes its books in a variety of electronic formats. Some content that appears in print may not be available in electronic format. For information about Apple Academic Press products, visit our website at **www.appleacademicpress.com** and the CRC Press website at **www.crcpress.com**

ABOUT THE EDITORS

Md. Hedayetullah, PhD

Md. Hedayetullah, PhD, is an Assistant Professor/Scientist and Officer In-Charge, AICRP (*All India Coordinated Research Projects*) on Chickpea, Directorate of Research, Bidhan Chandra Krishi Viswavidyalaya, Kalyani, Nadia, West Bengal. He is a former Agronomist with the NABARD, Balasore, Odisha, India. He was also formerly a Professor at the M.S. Swaminathan Institute of Agriculture Science, Centurion University of Technology and Management, Gajapati, Odisha, India, and an Assistant Professor at the College of Agriculture, Tripura, Government of Tripura, India. Dr. Hedayetullah is the author and co-author of 20 research papers, five review papers two book chapters, and one book.

Dr. Hedayetullah acquired his BS degree (Agriculture) from H.N.B. Garhwal University, Uttarakhand, India. He received his MS degree (Agronomy) from Palli Siksha Bhavana, Institute of Agriculture, Visva Bharati University, Sriniketan, West Bengal, India. He received his PhD (Agronomy) from Bidhan Chandra Krishi Viswavidyalaya, Mohanpur, Nadia, West Bengal, India. He was awarded the Maulana Azad National Fellowship Award from the University Grant Commission, New Delhi, India. He has received several fellowship grants from various funding agencies to carry out his research works during his academic career.

Parveen Zaman, PhD

Parveen Zaman, PhD, is an Assistant Director of Agriculture (Farm) at the Pulse & Oilseed Research Sub-Station, Beldanga, Department of Agriculture, Government of West Bengal, India. She is author and co-author of four research papers, three review papers, and four book chapters. She acquired her BS degree (Agriculture), receiving a Gold Medal, from Bidhan Chandra Krishi Viswavidyalaya, Mohanpur, Nadia, West Bengal, India. She received MS degree (Agronomy), also with Gold Medal, from Bidhan Chandra Krishi Viswavidyalaya, Mohanpur,

Nadia, West Bengal, India. She was awarded the Maulana Azad National Fellowship Award from the University Grants Commission, New Delhi, India, and has received several fellowship grants from various funding agencies to carry out her research works during her academic career.

CONTENTS

List of Contributors .... *xi*

List of Abbreviations .... *xv*

Acknowledgments .... *xvii*

Preface .... *xix*

PART I: Nonleguminous Perennial Forages **1**

1. Setaria Grass (African Grass) **3**
Md. Hedayetullah, Parveen Zaman, and Dhiman Mukherjee

2. Para Grass (Buffalo Grass) **13**
Md. Riton Chowdhury

3. Doob Grass (Bermuda Grass) **21**
Md. Hedayetullah and Parveen Zaman

4. Marvel Grass (Diaz Blue Stem) **31**
Golam Moinuddin

5. Pangola Grass (Digit or Woolly Finger Grass) **39**
Md. Hedayetullah, Parveen Zaman, and Rajib Kundu

6. Canary Grass (Harding Grass) **47**
Bulbul Ahmed and Md. Hedayetullah

7. Anjan Grass (African Foxtail Grass) **55**
Md. Hedayetullah, B. Bhattacharya, and Parveen Zaman

8. Rhodes Grass (Abyssinian Rhodes Grass) **65**
Md. Hedayetullah and Sibajee Banerjee

PART II: Leguminous Forages **75**

9. Grass Pea (Indian Vetch) **77**
Raghunath Sadhukhan, Md. Hedayetullah, and Parveen Zaman

10. Moth Bean (Dew Bean) 91
Parveen Zaman and Md. Hedayetullah

11. Soybean (Bhatman) 101
Sagar Maitra

12. Fenugreek (Greek Clover) 121
Parveen Zaman, Md. Hedayetullah, and Kajal Sengupta

13. Senji (Sweet Clover) 131
Md. Hedayetullah, Parveen Zaman, and Raghunath Sadhukhan

14. Mung Bean (Green Gram) 139
Kajal Sengupta

15. Urd Bean (Black Gram) 155
Kajal Sengupta and Md. Hedayetullah

PART III: Nonleguminous and Nongraminaceous Forages 169

16. Sunflower (Sujyomukhi) 171
A. Zaman and Parveen Zaman

17. Brassicas 187
Utpal Giri, Soma Giri, Navendu Nair, Abhijit Saha, Sonali Biswas, Niladri Paul, M. K. Nanda, and Protit Bandyopadhyay

18. Turnip (Salgam) 207
Parveen Zaman and Md. Hedayetullah

19. Gajar (Carrot) 215
Mohammed Abdel Fattah and Md. Hedayetullah

20. Amaranthus (Pigweed) 225
G. C. Bora

PART IV: Leguminous Perennial Trees 239

21. Subabul (River Tamarind) 241
Sagarika Borah, Mokidul Islam, and T. Samajdar

22. Gliricidia (Quickstick) 251
Golam Moinuddin

23. Khejri (Prosopis) **259**
Raj Kumar, B. S. Khadda, A. K. Rai, J. K. Jadav, and Shakti Khajuria

24. Nonconventional Legumes Forage Crops **287**
Dhiman Mukherjee and Md. Hedayetullah

25. Azolla: An Unconventional Forage Crop **309**
Dulal Chandra Roy, Abhijit Saha, and Sonali Biswas

26. Disease Management of Nonleguminous and Nongraminaceous Forages **321**
Diganggana Talukdar and Utpal Dey

Index **339**

LIST OF CONTRIBUTORS

Bulbul Ahmed
Scientific Officer, Plant Physiology Division, Bangladesh Agricultural Research Institute (BARI), Joydebpur, Gazipur 1701, Bangladesh

Protit Bandyopadhyay
Department of Agronomy, Bidhan Chandra Krishi Viswavidyalaya, Mohanpur 741252, Nadia, West Bengal, India

Sibajee Banerjee
Department of Agronomy, Bidhan Chandra Krishi Viswavidyalaya, Mohanpur, Kalyani 741235, Nadia, West Bengal, India

B. Bhattacharya
Seed Farm, AB Block, Bidhan Chandra Krishi Viswavidyalaya, Kalyani 741235, Nadia, West Bengal, India

Sonali Biswas
Department of Agronomy, Bidhan Chandra Krishi Viswavidyalaya, Mohanpur 741252, Nadia, West Bengal, India

G. C. Bora
Department of Plant Breeding & Genetics, Assam Agricultural University, Jorhat 785013, Assam, India

Sagarika Borah
Subject Matter Specialist, Krishi Vigyan Kendra, ICAR Research Complex for NEH Region, West Garo Hills, Umroi Road, Umiam 793103, Meghalaya, India

Md. Riton Chowdhury
AICRP on Sesame and Niger, Institute of Agricultural Science, University of Calcutta, 52/2 Hazra Road, Kolkata 700019, West Bengal, India

Utpal Dey
Division of Crop Production, ICAR Research Complex for NEH Region, Umaim 793103, Meghalaya, India

Mohammed Abdel Fattah
Department of Horticulture, Faculty of Agriculture, Cairo University, Giza, Egypt

Soma Giri
Horticulture Division, Krishi Vigyan Kendra, Ashoknagar, North 24 Parganas 743223, West Bengal, India

Utpal Giri
Department of Agronomy, College of Agriculture, Lembucherra, West Tripura 799210, Tripura, India

Md. Hedayetullah
Assistant Professor & Scientist (Agronomy), AICRP on Chickpea, Directorate of Research, Bidhan Chandra Krishi Viswavidyalaya, Kalyani 741235, Nadia, West Bengal, India
Department of Agronomy, Bidhan Chandra Krishi Viswavidyalaya, Mohanpur, Kalyani 741235, Nadia, West Bengal, India

Mokidul Islam
SMS, Krishi Vigyan Kendra, Ri-Bhoi, ICAR Research Complex for NEH Region, Umroi Road, Umiam 793103, Meghalaya, India

J. K. Jadav
SMS, Krishi Vigyan Kendra Panchmahal (ICAR-Central Institute for Arid Horticulture), Godhra-Vadodara Highway, Vejalpur, Panchmahal 389340, Gujarat, India

B. S. Khadda
SMS, Krishi Vigyan Kendra Panchmahal (ICAR-Central Institute for Arid Horticulture), Godhra-Vadodara Highway, Vejalpur, Panchmahal 389340, Gujarat, India

Shakti Khajuria
SMS, Krishi Vigyan Kendra Panchmahal (ICAR-Central Institute for Arid Horticulture), Godhra-Vadodara Highway, Vejalpur, Panchmahal 389340, Gujarat, India

Raj Kumar
SMS, Krishi Vigyan Kendra Panchmahal (ICAR-Central Institute for Arid Horticulture), Godhra-Vadodara Highway, Vejalpur, Panchmahal 389340, Gujarat, India

Rajib Kundu
Department of Agronomy, Bidhan Chandra Krishi Viswavidyalaya, Mohanpur 741252, Nadia, West Bengal, India

Sagar Maitra
Department of Agronomy, M. S. Swaminathan School of Agriculture, Centurion University of Technology and Management, Paralakhemundi 761211, Odisha, India

Golam Moinuddin
Assistant Professor, Regional Research Station, Bidhan Chandra Krishi Viswavidyalaya, Jhargram, West Bengal, India

Dhiman Mukherjee
Department of Agronomy, Bidhan Chandra Krishi Viswavidyalaya, Mohanpur, Kalyani 741235, West Bengal, India
AICRP on Wheat and Barley, Directorate of Research, Bidhan Chandra Krishi Viswavidyalaya, Kalyani 741235, West Bengal, India

Navendu Nair
Department of Entomology, College of Agriculture, Lembucherra, West Tripura 799210, Tripura, India

M. K. Nanda
Department of Agricultural Meteorology and Physics, Bidhan Chandra Krishi Viswavidyalaya, Mohanpur 741252, Nadia, West Bengal, India

Niladri Paul
Department of Soil Science and Agricultural Chemistry, College of Agriculture, Lembucherra, West Tripura 799210, Tripura, India

A. K. Rai
Krishi Vigyan Kendra Panchmahal (ICAR-Central Institute for Arid Horticulture), Godhra-Vadodara Highway, Vejalpur, Panchmahal 389340, Gujarat, India

Dulal Chandra Roy
Department of ILFC, WBUAFS, Mohanpur 741252, Nadia, West Bengal, India

Raghunath Sadhukhan
Department of Genetics and Plant Breeding, Bidhan Chandra Krishi Viswavidyalaya, Mohanpur 741252, Nadia, West Bengal, India

Abhijit Saha
Department of Agronomy, College of Agriculture, Lembucherra, West Tripura 799210, Tripura, India

T. Samajdar
Krishi Vigyan Kendra, ICAR Research Complex for NEH Region, West Garo Hills, Umroi Road, Umiam 793103, Meghalaya, India

Kajal Sengupta
Department of Agronomy, Bidhan Chandra Krishi Viswavidyalaya, Mohanpur, Kalyani 741235, Nadia, West Bengal, India

Diganggana Talukdar
Department of Plant Pathology and Microbiology, College of Horticulture, Central Agricultural University, Ranipool 737135, Sikkim, India

A. Zaman
Department of Agronomy, M. S. Swaminathan School of Agriculture, Centurion University of Technology and Management, Paralakhemundi 761211, Odisha, India

Parveen Zaman
Assistant Director of Agriculture, Pulse and Oilseed Research Sub-station, Department of Agriculture, Government of West Bengal, Beldanga 742133, Murshidabad, West Bengal, India

LIST OF ABBREVIATIONS

AAU	Assam Agricultural University
ADF	acid detergent fiber
ANF	antinutritional factor
Bt	*Bacillus thuringiensis*
CF	crude fiber
CP	crude protein
DAS	days after sowing
DAT	days after transplanting
DHP	3-hydroxy-4(1H) pyridine
DM	dry matter
EC	emulsifiable concentrate
EE	ether extract
FYM	farmyard manure
GFY	green fodder yield
IPM	integrated pest management
MB	mold board
NAA	naphthalene acetic acid
NFE	nitrogen-free extract
NF	nitrogen-free extract
NOA	2-naphthoxyacetic acid
NPV	nuclear polyhedrosis virus
SMCO	S-methyl cysteine sulphoxide
SSP	single superphosphate
TA	total acid
TDN	total digestable nitrogen
WG	wettable granule

ACKNOWLEDGMENTS

First of all, I ascribe all glory to the gracious "Almighty Allah" from whom all blessings come. I would like to thank him for His blessing to write this book.

I express my grateful thanks to my beloved wife, Parveen, for her wholehearted assistance. I express our deep sense of regard to my Abba, Maa, Jiju, Mehebub, Ismat didi, Tuhina, whose provided kind cooperation and constant encouragement.

With a profound and unfading sense of gratitude, I wish to express our sincere thank to the Bidhan Chandra Krishi Viswavidyalaya, India, for providing me with the opportunity and facilities to execute such an exciting project and for supporting my research.

I convey special thanks to my colleagues and other research team members for their support and encouragement and for helping me in every step of the way to accomplish this venture.

I am grateful to Dr. Md. Wasim Siddiqui, Mr. Ashish Kumar, and Mr. Rakesh Kumar from Apple Academic Press for helping me to accomplish my dream of publishing this book series, *Forage Crops of the World.*

PREFACE

Fodder production depends on soil type, land capability, cropping pattern, climate, and socioeconomic conditions. Agricultural animals are normally fed on the fodder available from cultivated areas, supplemented to a small extent by harvested grasses and top feeds. The three major sources of fodder are crop residues, cultivated fodder, and fodder from trees, pastures, and grazing lands. Forage crops are essential for quality milk and meat production. The patterns and types of fodder crops vary as per geographical location. In many countries of the world, people are not paying adequate attention to the feed and fodder for livestock and dairy animals. In addition green fodder, hay and silage also are important factors for their health, milk, and meat production during lean periods. Wide forage diversity exists throughout the world. Cultivated land is gradually decreasing, and within that, land resources to meet the need for food and fodder production to feed the world is also decreasing. Moreover, land resource for major fodder production is limited. We have to manage well to grow fodder crops with limited resources. The major fodder crops that are most nutritious to the animals need to be adopted in our cropping systems. Dual purpose crops have to be grown in cropping systems so that food and fodder grown together can meet the demand significantly under sustainable agriculture. Most of the fodder crops also have the human food value. In this respect, best utilization of fodder crops has to be adopted.

PART I
Nonleguminous Perennial Forages

CHAPTER 1

SETARIA GRASS (AFRICAN GRASS)

MD. HEDAYETULLAH[1*], PARVEEN ZAMAN[2], and
DHIMAN MUKHERJEE[1]

[1]*Department of Agronomy, Bidhan Chandra Krishi Viswavidyalaya, Mohanpur, Kalyani 741235, Nadia, West Bengal, India*

[2]*Assistant Director of Agriculture, Pulse and Oilseed Research Sub-station, Department of Agriculture, Government of West Bengal, Beldanga, Murshidabad 742133, West Bengal, India*

[*]*Corresponding author. E-mail: heaye.bckv@gmail.com*

ABSTRACT

The *Setaria* grass is also called "golden Timothy." The name is derived from the Latin word *seta*, meaning "bristle" or "hair," which refers to the bristly spikelets. This grass has a wide range of adaptability; it is more superior to other lines in quality and green fodder production. This grass requires minimum management for its growth. The *Setaria* grass is native of tropical and subtropical Africa. The first cut can be taken after sowing or planting of 9–10 months. The subsequent cuts can be taken after every 5–8 weeks depending on the season and growth of the grass. The green herbage yield varies 400–700 q/ha. The herbage yield obtained less in summer compare to rainy season. The crude protein of *Setaria anceps* varies between 4.8% and 18.4%. Seed is produced over much of the growing season, with good commercial yields usually of 100 kg/ha.

BOTANICAL CLASSIFICATION

Kingdom: Plantae
Order: Poales

Family: Poaceae
Genus: *Setaria*
Species: *anceps*
Binomial name: *Setaria anceps* Staph. ex Massey

1.1 BOTANICAL NAME

Setaria anceps Staph. ex Massey.

1.2 COMMON NAME

Golden Timothy, foxtail grass, bristle grass, African grass, and South African pigeon grass.

1.3 INTRODUCTION

The *Setaria* grass is also called "golden Timothy." The name is derived from the Latin word *seta*, meaning "bristle" or "hair," which refers to the bristly spikelets (Quattrocchi, 2000). The genus includes over 100 species distributed in many tropical and temperate regions around the world (Aliscioni et al., 2004), and members are commonly known as foxtail or bristle grasses. This grass is popularly known as *Setaria sphacelata*. It is the cultivated *Setaria* grass known to agriculturists under the name of *Setaria sphacelata*. *Setaria anceps* differs from *Setaria sphacelata* mainly by its vegetative characteristics, folded and sharply keeled leaf sheath arranged in fan fashion. *Setaria anceps* together with other species forms so-called *Setaria sphacelata* complex. This grass has a wide range of adaptability; it is more superior to other lines in quality and green fodder production. This grass requires minimum management for growth.

1.4 ORIGIN AND DISTRIBUTION

The *Setaria* grass is native of tropical and subtropical Africa. It is widely cultivated in Africa, Australia, Taiwan, New Guinea, Philippine, Israel, and Brazil. It was first introduced to India in 1950. It has now spread to all over India in medium rainfall areas. In North India, it grows in irrigated conditions.

1.5 BOTANY OF PLANT

The grass is perennial in nature; tufted stems are erect and height is about 1–2 m. The stems are moderately thick and usually compressed in lower part of the stem. The leaves are about 40-cm long and 10–20-mm wide, green to dark in color; sharply keeled leaf sheath arranged in fan fashion at the stem bases. The panicle is dense and cylindrical about 10–30-cm long and orange and purple in color. The spikelets are two in numbers; the lower one is male or sterile and the upper one is bisexual. The grain is elliptic in shape, concave on the embryo side, and flat on the other side.

1.6 CLIMATIC REQUIREMENTS

The *Setaria* grass occurs naturally in many countries such as Africa and Australia at 600–2600-m altitude. The *Setaria* grass usually grows under an annual rainfall of over 750 mm but in South Africa, some of its varieties exist even with an annual rainfall ranging from 500 to 750 mm. It also can grow vigorously under high rainfall areas between 1000 and 1500 mm. It cannot survive under hot and dry areas if it is extended for longer period. The grass grows very well at 20–25°C temperature. It also prefers a relative humidity from 70% to 80% for its rapid growth and it tolerates light frosts and fog. The *Setaria* grass is more cold tolerant than most of the other tropical and subtropical grasses. Under prolong hot and dry weather, the plants remain dormant (Chatterjee and Das, 1989).

1.7 SOIL REQUIREMENTS

Most commonly, it is found on soils with texture ranging from sand to clay loam and light clay, but it can be grown on heavy clay soil. It survives low fertility conditions but responds to improved fertility. It is not well adapted to alkaline or very acid soils, most wild collections coming from soils of pH 5.5–6.5. *Setaria* is generally low salt tolerant.

1.8 LAND PREPARATION

The perennial grass requires thorough cultivation and land preparation which may need one deep plowing with disk plow followed by 3–4 operations with a country plow or disk harrow. The land should be free from weeds.

1.9 SOWING

Planting of rooted slips can be done any time between February and November. In irrigated areas, it can start in the month of March. Under rainfed conditions, planting is done between June and August. The seedlings are raised in nursery and planted finally in the main field. The direct sowing also is done.

1.10 SPACING

The recommended spacing is 50 cm × 30 cm, but the row-to-row distance may be increased to 60–70 cm when the soil is poor and irrigation facility is not available. For intercropping with legumes, the row-to-row distance is maintained at 100 cm × 30 cm.

1.11 SEED RATE

The seed rate varies from 3.5 to 4 kg/ha if it is clean seed. In case of rooted slips, the planting rates vary between 800 and 1600 kg/ha or 33,000–67,000 rooted slips per hectare.

1.12 CROP MIXTURES

Setaria belongs to the Poaceae family, since it is not suitable combination to grow with other grasses. Number of pasture legumes can grow with *Setaria anceps*. It can be grown with lucerne, berseem, shftal, senji, metha, etc. in the winter and with cowpea, guar, rice bean, green gram, black gram, velvet bean, etc. in the summer. Legumes grown in mixtures with the grass seldom increase the yield of grass but they usually increase total dry matter and especially crude protein (CP) yields of the mixed herbage.

1.13 VARIETIES

1.13.1 NANDI SETARIA

Nandi Setaria has a variety of *Setaria anceps* and it is a native of tropical Africa extending from the north of the republic South Africa to Kenya. It originates in Nandi high lands, above 1200-m altitude of Kenya. *Nandi Setaria* does not thrive well in acidic soils. The grass has relatively low oxalate contents.

1.13.2 NANDI MARK 2

It is developed from single mass selection. It is more uniform and leafy. *Nandi mark* 3 is a further selection of Mark 2.

1.13.3 KAZUNGULA

It is the tall and vigorous plants with thicker stems, coarser, broader, and bluish green leaves. It is moderately drought tolerant.

1.13.4 NAROK SETARIA

It is more similar to the Nandi variety. It is more robust and vigorous, but not as coarse as Kazungula. It has good nutritive value and it is recommended for subtropical and high-altitude tropical areas with cool winters. It is developed in Australia and it is a tetraploid with 2n = 36. Some other varieties are most common in Australia such as Solander and Splenda.

1.14 MANURES AND FERTILIZERS APPLICATION

Setaria anceps responds well to nitrogenous fertilizers and produces over 30 kg dry matter or even 65 kg dry matter/kg of applied N. In natural grasslands, nitrogenous fertilizers respond well; *S. anceps* increases their herbage production. Because it grows best in at least moderately fertile soils, it is advisable to use an establishment application of 200–300 kg/ha of superphosphate and 50 kg/ha of muriate of potash on less fertile soils. Nitrogenous fertilizers like urea application as a basal at the rate of 100 kg/ha (46 kg/ha N) are beneficial in pure stands.

1.15 WATER MANAGEMENT

The grass prefers moist soil but not wet soil. The field should be drained well in rainy season if field is under submergence. For better crop establishment, presowing and postsowing irrigation is important. Subsequent irrigations should be given as and when necessary.

1.16 INTERCULTURAL OPERATION AND WEED MANAGEMENT

In the first 2–3 months of establishment, the crop requires one or two weeding and interculture operation. Every year of the cool season, interculture operation done country plow or hoe is necessary to control grassy, broad leaf, and sedge weeds and encourages the plant growth. *Setaria* generally competes aggressively with weeds. A strong stand of *Setaria* will reduce the frequency of fireweed (*Senecio madagascariensis* and *Senecio lautus*) to scarce levels. Established *Setaria* is tolerant of the broadleaf weed control herbicides 2,4-D, dicamba, and MCPA—check registration status and label instructions before use.

1.17 PEST AND DISEASES

Generally, no serious pest and diseases occur in *Setaria* grass. Attack by armyworm and pasture webworm can destroy much of the leaf, particularly young leaf. Buffel grass seed caterpillar can cause considerable damage to seed crops from late January onward. *Pyricularia* leaf spot is prevalent under hot humid conditions and can retard the growth of ungrazed stands.

1.18 HARVESTING

The first cut can be taken after sowing or planting of 9–10 months. The subsequent cuts can be taken after every 5–8 weeks depending on the season and growth of the grass. Generally, 7–8 cuts can be taken per year from the same field best management practices. After each cutting, a stubble height of 8–10 cm is left for good regeneration.

1.19 GREEN FODDER YIELD

S. anceps develops large number of tillers per stubble. The average tillers vary between 1400 and 1900 tillers/m^2. Nandi variety gives more tillers compare to other variety. The green herbage yield varies 400–700 q/ha. The herbage yield obtained less in summer compare to rainy season.

1.20 TOXICITIES

The oxalic acid toxicities are more common in *Setaria* species. The oxalic acid content in this grass is ranging between 2.78% and 7.13%. The high concentration of oxalic acid may cause serious health hazard to the livestock. The oxalate may accumulate in the kidneys to toxic levels resulting in death of the dairy animals and livestock. The sheep are faced such common problems in pastures. Pasture animals should give dicalcium phosphate in small quantities (15–20 g) in their salt ration to minimize the toxic hazards. *Setaria* develops high levels of oxalate in the leaf, especially in young, well-fertilized, vigorous growth. This causes hyperparathyroidism ("big head" disease) in horses and donkeys and can cause nephrosis (kidney disease) in ruminants.

It can also lead to hypocalcemia (milk fever) and hypomagnesemia (grass tetany) in ruminants, particularly in high-producing dairy cows. This is less of a problem in animals that graze *Setaria* regularly. "Kazungula" and "Splenda" develop the highest oxalate levels, followed by "Narok" and "Solander," with "Nandi" the lowest.

1.21 NUTRITIVE VALUE

The CP of *S. anceps* varies between 4.8% and 18.4%, digestible crude protein content from 2.1% to 13.2%, crude fiber (CF) content from 24% to 34.4%, and ether extract from 2.4% to 4.70%. The digestibility for CP varies between 44% and 72%, for CF from 65% to 77%, and for nitrogen-free extract from 40% to 50%.

1.22 PALATABILITY

Setaria is extremely palatable when young but becomes stemmy and unacceptable with onset of seeding due to loss of moisture from the stem and age.

1.23 UTILIZATIONS

The grass can be used as soilage, silage, and hay. The grass gives satisfactory silage with molasses. Acetic acid fermentation occurs but not lactic acid during fermentation. Good hay can be prepared from this grass and dry matter and nitrogen did not exceed 10%.

1.24 LIMITATIONS

High levels of oxalate in some varieties can cause problems with milking cows, horses, and donkeys. Quality drops rapidly with onset of seeding.

1.25 SEED PRODUCTION

Seed is produced over much of the growing season, with good commercial yields usually of 100 kg/ha. Seed crops are fertilized with 100–150 kg/ha of N after a cleaning and are usually direct headed when 10–15% of the seed has shattered.

FIGURE 1.1 **(See color insert.)** Setaria grass.

KEYWORDS

- *Setaria* grass
- agronomic package and practices
- green fodder yield
- nutritive value

REFERENCES

Aliscioni, S., et al. An Overview of the Genus *Setaria* in the Old World: Systematic Revision and Phytogenetic Approach. Abstract. *Botany.* Salt Lake City. August 3, 2004.

Chatterjee, B. N.; Das, P. K. *Forage Crop Production Principles and Practices*; Oxford & IBH Publishing Co. Pvt. Ltd.: Kolkata, 1989.

Quattrocchi, U. *CRC World Dictionary of Plant Names: Common Names, Scientific Names, Eponyms, Synonyms, and Etymology;* CRC Press: Boca Raton, FL, 2000; p 2470. ISBN 0-8493-2673-7.

CHAPTER 2

PARA GRASS (BUFFALO GRASS)

MD. RITON CHOWDHURY*

AICRP on Sesame and Niger, Institute of Agricultural Science, University of Calcutta, 52/2 Hazra Road, Kolkata 700019, West Bengal, India

**E-mail: riton.85@hotmail.com*

ABSTRACT

Para grass is a vigorous perennial grass used as fodder for the animal and livestock. It is also known as buffalo grass, Mauritius signal grass, California grass, giant couch grass, water grass, and pani wali ghas. Para grass is grown widely in marshy lands as used as fodder. The grass is used as green fodder, soiling, and even for dry hay. The stems have hairy nodes. Average green fodder yield is about 800–1000 q/ha. Under favorable condition, green fodder yield can be increased up to 1200–1500 q/ha. This grass is highly palatable and nutritious. It contains an average 10.2% crude protein and 23.6% crude fiber in fresh grass.

BOTANICAL CLASSIFICATION

Kingdom: Plantae
Order: Poales
Family: Poaceae
Genus: *Brachiaria*
Species: *mutica*
Binomial name: *Brachiaria mutica*

2.1 BOTANICAL NAME

Brachiaria mutica (Forssk.) Stapf.

2.2 COMMON NAME

Buffalo grass, Mauritius signal grass, pasto pare, malojilla, gramalote, parana, Carib grass, Scotch grass, and California grass.

2.3 SYNONYMOUS

Brachiaria numidiana (Lam.) Henrard; *Brachiaria purpurascens* (Raddi) Henrard; *Panicum amphibium* Steud.; *Panicum barbinode* Trin.; *Panicum equinum* Salzm. ex Steud.; *Panicum molle* Sw.; *Panicum muticum* Forssk.; *Panicum numidianum* Lam.; *Panicum pictigluma* Steud.; *Panicum punctatum* Burm. f.; *Panicum punctulatum* Arn. ex Steud.; *Panicum purpurascens* Raddi; *Panicum sarmentosum* Benth.; *Paspalidium punctatum* (Burm. f.) A. Camus; *Setaria punctata* (Burm. f.) Veldkamp; *Urochloa mutica* (Forssk.) Nguyen; *Urochloa mutica* (Forssk.).

2.4 INTRODUCTION

Para grass is a vigorous, semiprostrate perennial grass used as fodder for the cattle. It also commonly known as buffalo grass, Mauritius signal grass, California grass, Giant couch grass, water grass, and pani wali ghas. Para grass is grown widely in marshy lands as used as fodder. It grows in wide range of soil. It is best grown on the submerged or low-lying areas as well as on saline soils where nothing else survives. Para grass favors hot and humid climate of tropics and subtropics with high annual rainfall ranging between 900 and 1000 mm. This grass is highly palatable and nutritious (ICAR, 2012). It contains an average 10.2% crude protein and 23.6% crude fiber in fresh grass. The grass is used as green fodder, soiling, and even for dry hay.

2.5 ORIGIN AND DISTRIBUTION

Para grass is probably originated in Brazil, South America. Some documents also support that it was probably native to flood plains of sub-Saharan tropical Africa. Later on, it was introduced to Australia, and from there, it spread to the South Asian countries. It was introduced to India in 1894 at Pune from Sri Lanka. Now, it is widely grown in the tropical regions of South and Central America, Africa, Bangladesh, India, Sri Lanka, Australia, and Southeast Asia. In India, it is grown in Tamil Nadu, Odisha, West Bengal, Kerala, Karnataka, Bihar, Assam, and some parts of Maharashtra and Madhya Pradesh.

2.6 PLANT CHARACTERISTICS

Para grass is a vigorous, semiprostrate perennial grass with creeping stolons which can grow up to 5-m (16 ft) long. The stems have hairy nodes and leaf sheaths and the leaf blades are up to 2-cm (0.8 in.) wide and 30-cm (12 in.) long. It roots at the nodes and detached pieces of the plant will easily take root in moist ground. The flowerhead is a loose panicle up to 30-cm (12 in.) long with spreading branches. The paired spikelets are arranged in uneven rows and are elliptical and 2.5–5-mm (0.1–0.2 in.) long. The rachis is tinged with purple. Although many flowerheads grow, only a few viable seeds are produced, and propagation is usually by vegetative means. Para grass can be distinguished from the closely related tanner grass by its paired spikelet, tanner grass having single spikelets. The stems are 1–2 m in height with profuse rooting at nodes having forming dense cover with ascending branches. The culms are hollow, succulent with hairy internodes. The leaf blade is dark green in color and is 25–30-cm long and 1–2 cm in width. Inflorescence is open panicle and about 10–20-cm long with solitary racemose, acute, irregularly multiheritage spikelet 3–3.5-mm long. It is a short day plant (Chatterjee, 1989).

2.7 CLIMATIC REQUIREMENTS

Para grass favors hot and humid climate of tropics and subtropics with high annual rainfall ranging between 900 and 1000 mm. The crop can withstand short flooding and waterlogging. It cannot be grown in dryland and in arid and semiarid regions. It grows well in waterlogging conditions on river and canal banks and can withstand prolonged flooding. The optimum temperature for growth and development is 25°C. The crop can be grown through the year in the regions where temperature remains above 15°C. This grass is more suited for water inundated conditions and sewage farms (Chatterjee, 1989). It is sensitive to cold (temperature below 15°C) and makes little or no growth during winter months of December–February in subtropical regions of India.

2.8 SOIL REQUIREMENTS

It grows in wide range of soils. It is best grown on the submerged or low lying areas as well as on saline soils where nothing else survives. It tolerates slightly acid to alkaline soils; that is why, it is an excellent grass in soil reclamation. It performs well in moist sewage farm soils, low laying areas beside rivers or canals, and soils too wet for normal crops.

It can be grown even on sandy soils with good irrigation facility. Heavy textured clay soil with high water holding capacity is considered to be best for para grass. Experimental results suggest that addition of pond sediment into soil increases the productivity.

2.9 LAND PREPARATION

The land should be plowed properly and well pulverized by two to three crosswise plowing followed by laddering. The weeds should be removed properly before planting the slips.

2.10 SEED RATE AND SOWING METHOD

Under irrigated conditions, the best time of planting is in March while under rainfed conditions at the onset of monsoon. In South Indian conditions, it can be planted throughout the years. Seed setting is a major problem for this crop, so thin shoot is used as planting material. Shoot slips of 30-cm length each with two to three nodes are best for planting. Line sowing is mostly preferred with a distance about 50–60 cm between rows and 20–40 cm between plants. To save labor cost, scattering of rooted slips and crosswise plowing is practiced mostly for monsoon crop, after 3–4-year rejuvenation of old pasture is done to promote fresh growth. On average, about 27,000–40,000 slips (8–10 q) are required to plant 1 ha of land. Seeds can also be used for direct sowing or sowing in nursery for transplantation. But, poor seed setting is usually discouraged for propagation by seeds. The recommended seed rate is 2.5–3.5 kg/ha.

2.11 CROP MIXTURE

This crop performs well with crops such as *Vicia*, *Lotononis* sp., Stylo, berseem, senji, and *Calopogonium* as crop mixture but performs best as pure crop. The vigorous growth of the crop subpresses other crops' growth. Cultivation with legumes helps to improve the soil fertility. Though some intercropping systems with legumes are suggested, it is widely grown as a pure crop (Amam et al., 2015).

2.12 NUTRIENT MANAGEMENT

Wastewater from cattle shed or any swage water is good for the crop. In general, 10–15-t well-prepared farmyard manure or compost per hectare is applied 25–30 days prior to sowing. Being a fodder crop it responds well under split application of nitrogen. During final land preparation, 150–200 kg of nitrogen per hectare is applied as basal dose. After each cutting, 25–30 kg of N ha^{-1} is applied for a better growth and improves protein content. In each year, 50–60 kg of P_2O_5 along with 50 kg ha^{-1} of potassium is to be applied as basal. This dose is to be repeated every year on the onset of monsoon or in the month of June–July. Application of phosphorous is very important for a good root growth and to maintain a favorable shoot:root ratio (Mukherjee et al., 2008).

2.13 WATER MANAGEMENT

Mostly, the crop is grown as a rainfed crop in our country. Irrigation should be given at equally close intervals as per numbers of cutting taken under irrigated condition. During initial establishment, phase two to three light irrigation is necessary to apply. Later on, irrigation at 10–15-day intervals is profitable during summer. Water requirement of the crop is low but the land should always be kept moist throughout its growing period. The crop can withstand short-term waterlogging.

2.14 VARIETIES

In India, there are no improved varieties of this grass yet available; only local varieties are cultivated.

2.15 WEED MANAGEMENT

Weed control is not so important because it covers the ground very quickly which reduces the weed incidence. During early establishment, phase one or two hand weeding may be given with proper care so that the growing runners will not damage. So, one preemergence application of herbicide such as pendimethalin @ 0.75–1 kg a.i./ha is recommended at 1–3 days after transplanting (DAT) to control the initial flush of weeds because this first 2 months of establishment phase is more sensitive to weed. The weeds appear later on suppressed by the vigorous growth of the crop.

2.16 INSECT-PEST AND DISEASES

In India, no such incident has reported for pest and diseases. Under prolonged waterlogged condition followed by humid condition, sheath blight occurred in few cases. Incidence of few insects such as common armyworm (*Mythimna convecta*) and day feeding army worm (*Spodoptera exempta*) were reported form Queensland, Australia. The leaf hopper (*Toya* sp.) was found to be the most destructive one (Mukherjee et al., 2008). The incidence of mold fungus infestation on the tender leaves was (*Capnodium* sp.) also reported from parts of South America.

2.17 HARVESTING

In the early establishment stage, the grass spreads its runner and roots developed from the nodes and erect shoots were produced. This procedure takes a few days to cover the whole field. Thus, the first cut delays. Total six to eight cuts can be taken in a year. The first cutting is taken 70–80 days after planting and the subsequent cuts at 35–45-day intervals. A cut may take when the plant's height is 60–70 cm.

2.18 GREEN FODDER YIELD

Average green fodder yield is about 800–1000 q ha^{-1}. Under favorable condition, green fodder yield can be increased up to 1200–1500 q ha^{-1}. In each cutting on an average, 150 q/ha green fodder can be obtained. The average yield for North Indian condition is 750 q/ha, whereas in South India, a highest yield of 1300 q ha^{-1} can be obtained (Mukherjee et al., 2008).

2.19 NUTRITIVE VALUE

The crop contains 2.8–16.1% of crude protein, 0.32–0.76% calcium, and 0.35–0.8% phosphorus with 28–34% crude fiber. It also contains 41–57% nitrogen-free extract (Jones and Csurhes, 2012).

2.20 USE

Para grass is nontoxic in nature. This grass is fed in the green form. Young leaves and stems are palatable. It is not popular for conservation either as hay or as silage because with age, leaves, and stems become coarse. The green harvest contains higher quantity of lipid and protein contain than rice straw so it can be a good substitute.

KEYWORDS

- **para grass**
- **agronomic management**
- **green fodder yield**
- **nutritive value**

REFERENCES

Amam, M. R.; Haque, M. M.; Sumi, K. R.; Ali, M. M. Proximate Composition of Paragrass (*Brachiaria mutica*) Produced in Integrated Fish–Fodder Culture System. *Bang. J. Anim. Sci.* **2015,** *44* (2), 113–119.

Chatterjee, B. N. *Forage Crop Production—Principles and Practices*; Oxford & IBH: Kolkata, 1989.

Indian Council for Agricultural Research (ICAR). *Handbook of Agriculture*, 6th ed.; Indian Council for Agricultural Research (ICAR): New Delhi, 2012.

Jones, M. H.; Csurhes, S. *Invasive Species Risk Assessment: Para Grass* (*Urochloa mutica*); Queensland Department of Agriculture, Fisheries and Forestry: Mackay, 2012.

Mukherjee, A. K.; Maiti, S.; Mandal, S. S. *Forage Production Technology Manual*; Deptartment of Agronomy, BCKV: Mohanpur, 2008.

CHAPTER 3

DOOB GRASS (BERMUDA GRASS)

MD. HEDAYETULLAH[1*] and PARVEEN ZAMAN[2]

[1]*Department of Agronomy, Bidhan Chandra Krishi Viswavidyalaya, Mohanpur, Kalyani 741235, Nadia, West Bengal, India*

[2]*Assistant Director of Agriculture, Pulse and Oilseed Research Sub-station, Department of Agriculture, Government of West Bengal, Beldanga 742133, Murshidabad, West Bengal, India*

[*]*Corresponding author. E-mail: heaye.bckv@gmail.com*

ABSTRACT

Doob grass is also known as Bahama grass, dog's tooth grass, devil's grass, couch grass, wire grass, and scutch grass. Doob grass have high nutritional value, excellent palatability, persistence even under adverse conditions, high tolcrance to intensive grazing, and unique soil-binding capacity for soil conservation. Bermuda grasses establish rapidly and spread by vegetative propagules, both above ground (stolons) and below ground (rhizomes). It grows well on alluvial, red, and black soils with pH ranging from 5.5 to 8.0. The grass can tolerate considerable amount of salinity and alkalinity. The first cut is usually taken about 90–100 days after planting or sowing. Clipping is done at 4–5 cm above the ground level. The average green herbage yields from each cut are about 70–80 q/ha. The monsoon yields are higher than those of the hot and dry season. The average annual green herbage yields are 300 q/ha in North India and 450 q/ha in South India. It is palatable and acceptable to all types of livestock and dairy animals. The dry matter (DM) content is about 25–30% of fresh herbage. Crude protein content ranges from 7% to 18% of the DM.

3.1 BOTANICAL CLASSIFICATION

Kingdom: Plantae
Order: Poales
Family: Poaceae
Genus: *Cynodon*
Species: *dactylon* (L.) Pers
Binomial name: *Cynodon dactylon* (L.) Pers

3.2 BOTANICAL NAME

Cynodon dactylon (Pers.) Stent (Synonymous: *Cynodan polevansii*).

3.3 COMMON NAME

Doob (Sanskrit—Durva, Hindi—Dhub, Tamil—Arugam pillu; Kannada—Garika bullu). It is also known as Bahama grass, dog's tooth grass, devil's grass, couch grass, wire grass, and scutch grass.

3.4 INTRODUCTION

Doob grasses have high nutritional value, excellent palatability, persistence even under adverse conditions, high tolerance to intensive grazing, and have unique soil-binding capacity for soil conservation. This species, which is native to Africa, produces a vigorous, low-growing turf grass stand with high density and tolerances to both traffic and drought stress. Bermuda grasses establish rapidly and spread by vegetative propagules, both above ground (stolons) and below ground (rhizomes). Doob grass is also known as Bahama, or Bermuda grass. Being a troublesome weed in the cultivated fields, it is also called devils grass. Because of the dog tooth-like sheath on the stolons, it is also known as dog tooth grass. The other name of this grass is star grass, lawn grass, wire grass, and couch grass. Cynodon means dog tooth and dactylas mean finger like spikes, so the grasses are called *Cynodon dactylon*.

3.5 DESCRIPTION

Doob is one of the most controversial grasses. Bermuda grass has high nutritional value, excellent palatability, persistence even under adverse conditions, high tolerance to intensive grazing, unique soil-binding capacity for soil conservation, cosmopolitan nature with regard to texture, pH, and moisture content of the soil, while on the other hand, it is one of the most noxious weeds in arable farming.

3.6 PLANT CHARACTERISTICS

C. dactylon is a cross-pollinated species. Its variability is wide. The plants differ in yields, leafiness, response to fertilizers, etc. It is a stoloniferous and rhizomatous perennial with slender to stout stems. Leaves are flat or folded, 3–12-cm long and 2–4-cm wide. Ligule is a conspicuous ring of white hairs. Spikes are three to six, slender, 2–5-cm long; digitately arranged in single whorl. Spikelets are one flowered, awnless, laterally compressed, and arranged on one side of the rachis. The grass is very aggressive.

3.7 ORIGIN AND DISTRIBUTION

Doob grass is found in a tropical, subtropical, and warm temperate region of the world. It is a native of India or East Africa or Middle East (Farsani et al., 2012). Although it is not native to Bermuda, in Bermuda it is known as crab grass. It is ubiquitous, occurring in heavily grazed grasslands, roadsides, fallows and as a weed of arable land. In India, it is found in the grass covers of *Dicanthium* or *Cenchrus* or *Elyonurus* and *Schimal* or *Dicanthium* types.

3.8 CLIMATIC REQUIREMENTS

C. dactylon is widely cultivated in warm climates all over the world between about 30° S and 30° N latitude. The grass is sensitive to light intensity. High air temperatures, about 37°C, are required for maximum photosynthetic activity. Doob grows throughout the year in southern and eastern India where the temperatures do not drop below 15°C. It occurs even up to elevations of 2130 m. It is best suited to the tropical belt with 1000–1300 mm annual rainfall. High humidity and intermittent showers are most congenial for its growth. In North India, its growth is retarded in the winter months. Under dry conditions, the growth is less and it can remain dormant for 6–7 months.

It grows luxuriantly with the monsoon rains from June to September in the uplands where water does not stagnate for long. Well-drained, fertile clay to clay loam soils are the best but it can grow even on sandy loam soils. It grows well on alluvial, red, and black soils with pH ranging from 5.5 to 8.0. The grass can tolerate considerable amount of salinity and alkalinity (Chatterjee and Das, 1989).

3.9 SOIL AND ITS PREPARATION

All Bermuda grasses tolerate a wide range of soil types. A few light and shallow cultivations are enough to uproot the weeds and natural grasses from the field. Sometimes, deep plows may be necessary to bury the weeds. Seed crops require fine tilth and weed-free plots.

3.10 VARIETIES

C. dactylon is a variable species. The following are the important varieties of the tropics:

1. Var. *dactylon*, a tetraploid, grows on alkaline and saline soils and tolerates droughts but is absent in arid areas. It is highly valued as a quality pasture grass and it tolerates heavy grazing. Tift Bermuda was an improved selection of the variety in Georgia. The variety has strong, stout, and much branched rhizomes which spread readily and make eradication difficult. A course and robust race is also common in the Mediterranean area and southern Russia.
2. Var. *aridus* Harlon and de Wilt, a diploid, varies in habit. A small and unproductive type occurs in India while the east African type is large, robust, and vigorous.

3. Var. *elegans* Rendle, a tetraploid, occurs in South Africa. It has much branched stolons, rhizomes, and slender but wiry stems. The plants form a good grazing cover with dense low swards up to 30 cm in height. It also makes a good lawn. As a tropical turf grass, the variety is known by Cape Royal and Maadi River grasses.
4. Var. *coursii* Harlan and se Wilt, a tetraploid is a nonrhizomatous grass in the Madagascar highlands.
5. In Hybrid cultivars, *Cynodon dactylon* var., *dactylon* has been a serious weed of arable crops. Hybrid Bermuda grass (cv Coastal from var. *dactylon* × var. *elegans*) is suitable for grazing and forage production cv. Coastal is almost seed sterile and produces few fertile stems and panicles. It is resistant to common diseases and root-knot nematodes. It propagates easily and yields one and a half to four times more herbage than the local types. It sharply increased production and revolutionized the livestock industry of the southern states.

3.11 SOWING/PLANTING TIME

In North India, sowing is taken up in July when soil moisture is enough. Summer sowing usually gives patchy germination. In southern and eastern India, sowing can be done all the year round provided moisture is there in the soil. The grass is established from seeds, stolons, or rhizomes. Hulled seeds germinate more quickly than unhulled seeds. The seeds may be sown by *kera* method or broadcast on the moist soil and covered by light harrowing. The seeds may be sown in nursery and transplanted in the main field after 40–50 days (Chatterjee and Das, 1989).

But doob being a shy seeder, stolon, or rhizome springs are planted either in hills or rows. After planting, the soil should be pressed well to put the springs in contact with moisture. The tips of the springs should remain free on the surface of the soil. During establishment, stolons spread first; then, the underground rhizomes develop. Land should be irrigated if soil dries up.

3.12 SEED TREATMENT AND SEED INOCULATION

Seeds should be treated with Thiram at the rate of 3 g per kilogram of seeds. Seeds should be inoculated with bacteria *Azotobacter* before sowing for fixation of atmospheric nitrogen. Apply 20 g *Azotobacter* for 1 kg of seed before sowing for nitrogen fixation.

3.13 SEED RATE AND SOWING METHOD

The plating rate of springs is 3–4 q/ha. Freshly cut moist root stocks should be used for planting. Planting is best done in drizzling weather for well establishment of springs. For hulled seeds, the seed rate ranges between 3 and 3.5 kg/ha; for unhulled seeds, it ranges between 4.5 and 5 kg/ha, or even more; and for making turf, it is about 60 kg/ha.

3.14 SPACING

The row-to-row spacing maintained for optimum forage production is 50–60 cm and springs-to-springs or seed-to-seed distance is maintained about 30 cm. The spacing between two hills may be as high as 110 cm.

3.15 CROPPING SYSTEMS

The doob grass is not usually grown in association with any leguminous fodder crops because this grass can produce adverse effect clover seed germination. In some tropical countries, *Centrosema* is profitably intercropped at the rate of 5–6 kg/ha. The grass is grown in widely spaced rows when *Leucaena leucocephala* is grown between the rows. In the Southern United States of America, *Trifilium incarnatum*, *vicia*, *villosa*, and *Trifolium repens* are frequent components of Bermuda grass mixtures. In Florida, *Arachis glabrata* is a good associate with the Bermuda grass.

3.16 NUTRIENT MANAGEMENT (MANURES AND FERTILIZERS)

At the time of land preparation, farmyard manure or compost is applied @10–15 t/ha for better plant growth and higher green forage yield, the rate of nitrogen is less recommended in dry areas than the high rainfall areas.

The application of 1 kg nitrogen can produce 30–35 kg dry matter (DM) production. Phosphorus and potash are applied at the rate of 100 kg P_2O_5 and 100 kg of K_2O. Sometimes S deficiency is reported in that case N and P_2O_5 can be applied in the form of ammonium sulfate and superphosphate. In barren sandy, soils a 4:1:2 ratio of N:K_2O is required for balanced fertilization. After each and every cut, the grass is to be top dressed with 25 kg N/ha. A natural pasture of the grass is required to be fertilized with 30 kg N and 25 kg P_2O_5 per hectare with onset of monsoon to increase the green biomass and forage quality.

3.17 WATER MANAGEMENT

Doob prefers sufficient soil moisture throughout the life period with high N. In dry season, the crop should be irrigated after every top dressing for better uptake of fertilizer nitrogen. During summer, crop should irrigate at 15 days interval and every fourth week in the postmonsoon season. For higher green fodder production, soil moisture is very important factor. Water stagnation does not prefer this crop. Actively growing Bermuda grasses require (on average) approximately 1–2 in. of water per week (Brosnan and Deputy, 2008).

3.18 INTERCULTURE OPERATION AND WEED MANAGEMENT

Intercultural operation may be done with country plough, harrow, or hoe in every 2–3 weeks if is necessary. To control broad leaf weed, chemical weed management may adopted with application of 2,4-D. Grazing and cutting at early stage is also recommended for controlling of weeds. After 2–3 years, weeds can be destroyed and the grass is renovated by light disking, harrowing, and plowing.

3.19 INSECT-PEST AND DISEASE MANAGEMENT

Insect-pest and diseases are not well known in doob grass. For controlling insect-pest and diseases, the chemical management procedures are generally followed.

3.20 HARVESTING

The first cut is usually taken about 90–100 days after planting or sowing. Clipping is done at 4–5 cm above the ground level. The new growth comes from the adventitious buds near the soil surface. In North India, four cuttings are taken while in the Southern and Eastern India, five to six cuts are possible. Clipping may be done when the grass attains a height of 15–20 cm. After three to four times grazing, the grass should be clipped close to the ground to control weeds. The grass withstands trampling in grazing.

3.21 GREEN FODDER YIELD

The average green herbage yields from each cut are about 70–80 q/ha. The monsoon yields are higher than those of the hot and dry season. The average annual green herbage yields are 300 q/ha in North and 450 q/ha in South India.

3.22 SEED PRODUCTION

Common Bermuda grasses can be established from seed. While this may be cheaper than vegetative propagation, many seeded Bermuda grasses do not provide the same level of quality as hybrid Bermuda grass cultivars (Brosnan and Deputy, 2008). *C. dactylon* is shy in seed production. In drier areas, two seed crops may be harvested every year—one in April and May and the other in November and December when 150–200 kg of seed may be obtained per harvest per hectare.

3.23 NUTRITIVE VALUE

C. dactylon is one of the nutritious grasses among all grasses. It is palatable and acceptable to all types of livestock and dairy animals. The DM content is about 25–30% of fresh herbage. Crude protein (CP) content ranges from 7% to 18% of the DM and generally higher than many other tropical and subtropical grasses. CP content gradually decreases with plant age and the leaf: stem ratio is high. The digestibility for DM ranges between 45% and 60%, CP between 50% and 77%, crude fiber between 25% and 35% and nitrogen-free extract between 50% and 60%.

3.24 UTILIZATIONS

Excellent hay is prepared and fed to the animals. It is quickly dried in the sun and stored in well ventilized rooms. The artificially dried grass is used as hay and sometimes it is pelleted to grain better live weight, but it does not make good silage. It is used for lawns and turfs and on embankments as a soil binder for erosion control.

3.25 SPECIAL FEATURES (TOXICITIES)

The doob grass is free from hydrocyanic acid toxicity.

FIGURE 3.1 (See color insert.) Bermuda grass.

KEYWORDS

- **Bermuda grass**
- **agronomic package and practices**
- **fodder yield**
- **quality**
- **herbage**

REFERENCES

Brosnan, J. T.; Deputy, J. Bermuda Grass. *Turf Manage.* **2008,** *5*, 1–6.

Chatterjee, B. N.; Das, P. K. *Forage Crop Production Principles and Practices*; Oxford & IBH Publishing Co. Pvt. Ltd.: Kolkata, 1989.

Farsani, T. M.; Etemadi, N.; Sayed-tabatabaei, B. E.; Talebi, M. Assessment of Genetic Diversity of Bermudagrass (*Cynodon dactylon*) Using ISSR Markers. *Int. J. Mol. Sci.* **2012,** *13* (1), 383–392.

CHAPTER 4

MARVEL GRASS (DIAZ BLUE STEM)

GOLAM MOINUDDIN*

Regional Research Station, Bidhan Chandra Krishi Viswavidyalaya, Jhargram, West Bengal, India

**E-mail: moinuddin777@rediffmail.com*

ABSTRACT

Marvel grass is a popular forage crop and very much palatable for ruminants. Marvel grass originates from North Africa and India. It was introduced to Southern Africa, tropical America, the Caribbean, Southeast Asia, China, the Pacific Islands, and Australia. This is a popular pasture grass in many areas. It can be used in fields for grazing livestock and cut for hay and silage. Marvel grasses yield about 15–60 q/ha of dry matter; however under irrigated condition, the yield may go up to 170 q/ha. Green forage production ranges from 60 to 100 q/ha. The grass may produce 40–90 kg seed/ha. The grass contains 2.6–10.4% of crude protein, 34.9–45.5% crude fiber, and 7.1% lignin.

4.1 BOTANICAL CLASSIFICATION

Kingdom: Plantae
Order: Poales
Family: Poaceae
Subfamily: Panicoideae
Genus: *Dichanthium*
Species: *annulatum*
Binomial name: *Dichanthium annulatum* (Forssk.) Stapf

4.2 SYNONYMS

Andropogon annulatus, *Andropogon papillosus*, *Dichanthium nodosum*, *Dichanthium papillosum*.

4.3 COMMON NAME

There are so many common names of marvel grass for different parts of the world such as blue stem (the United States), Hindi grass, Delhi grass, Santa barbara grass (English), Sheda grass (Australia), Yerba de vias (Spanish), karad (India), pitilla (Cuba), etc.

4.4 BOTANICAL NAME

Dichanthium annulatum (Forssk.) Stapf.

4.5 INTRODUCTION

Marvel grass is an excellent and widely used fodder grass much appreciated by animal and livestock. Marvel grass is a popular forage crops and very much palatable for ruminants. Marvel grass originates from North Africa and India. It was introduced to Southern Africa, tropical America, the Caribbean, Southeast Asia, China, the Pacific Islands, and Australia. This is a popular pasture grass in many areas. It can be used in fields for grazing livestock and cut for hay and silage. It is tolerant of varied soil conditions, including soils high in clay and sand, poorly drained soils. In India, it is commonly grazed by sheep and goats. In case of dry forage, this grass is supplemented with some energy source.

4.6 ORIGIN AND DISTRIBUTION

Marvel grass is supposed to be originating from North Africa and India. Then, it was introduced into Southern Africa, tropical America, Pacific Islands, Northern Australia, Southeast Asia, China, and New Guinea. It is generally found within 8–28° in northern hemisphere up to an elevation of 600 m in dry to moist subtropical and tropical areas. In India, this grass occurs throughout the plains and hills at an elevation of 250–1375 m.

4.7 PLANT CHARACTERISTICS

It is a tufted perennial grass and grows up to a height of 60–100 cm. Plant is erect in nature with linear leaf blade of 5–25-cm length and 2–5-mm wide. Sometimes plant forms creeping stolon. Leaf sheaths are glabrous and ligules are of 1–1.8 mm in length. Most of the root is confined within 1 m of soil depth. The grass produces productive tillers of 1 m length. There is 3–5-mm long hair on nodes. Inflorescence consists of two to nine pale green or purple racemes. Spike consists of both sessile and pedicellate spikelets. Seed is 2 mm in length and oblong–ovate type.

4.8 CLIMATIC REQUIREMENTS

The grass is grown up to an altitude of 250–1300 m. It grows well both in tropical and subtropical areas having annual rainfall of 700–1400 mm. In low rainfall areas, low-lying areas are suitable for cultivation of this crop. It is fairly drought tolerant and optimum temperature for seed germination is 32–40°C. It does not prefer shade for optimum growth (Manidool, 1992).

4.9 SOIL AND ITS PREPARATION

The grass can be cultivated in a wide range of soil but it favors black cotton soil particularly in India. Acidic soil causes hazards for plant growth; however, it can withstand soil pH up to 5.5. It is very much tolerant to saline and alkaline soil. It requires fine seed bed for better crop stand but cultivation is also possible in rough seed bed. It is fairly tolerant to standing water (Bogdan, 1977).

4.10 VARIETIES

The important forage grass varieties are Marvel 8 (CPI 106073), Kleberg, T 587, PMT 587, Pretoria 90 (PI 188926, BN-6730, T-20090), IGFRI-S-495-1, and IGFRI-S-495-59 (lines selected at IGFRI, Jhansi).

4.11 SOWING TIME

The crop is generally cultivated during wet season from June to November particularly in India. It can also be cultivated from February to March.

4.12 SEED TREATMENT

It should be treated with fungicide such as carbendazim, Thiram, ziram, and captan or by *Trichoderma viride*.

4.13 SEED RATE AND SOWING METHOD

As seeds are not available commercially, it is generally cultivated by using root slips. Root slips are planted in line at a spacing of 60 cm × 60 cm. If seeds are used for sowing purpose, then seedlings are raised from seeds in nursery bed.

4.14 INTERCROPPING

Once crops get established, it becomes very aggressive; some grasses like *Dichanthium aristatum*, *Dichanthium caricosum*, legume plants such as *Medicago* sp., *Stylosanthes* sp. may compete successfully with marvel grass (Mehra et al., 1960).

4.15 CROP MIXTURE

This grass is not recommended for mixed cropping as it crowds out other grasses.

4.16 NUTRIENT MANAGEMENT

Generally, this crop does not require fertilizer; however, it responds to low-to-moderate level of nitrogen. However, a fertilizer schedule of 120–200 q/ha farmyard manure, 40–45 kg N/ha and 30–35 kg P_2O_5/ha, and 20 kg K_2O/ha per year can be maintained for good crop establishment and yield (Gill, 1970).

4.17 WATER MANAGEMENT

For wet season planning crop, generally it does not require irrigation, but one or two life-saving irrigation may be done as and when required.

4.18 WEED MANAGEMENT

Weed is not a serious problem for cultivation as it form tuft.

4.19 INSECT-PEST AND DISEASES

No serious pest and disease infestation is found; however, different types of fungi such as *Balansia sclerotica*, *Physoderma dichanthicola*, *Pithomyces graminicola*, *Puccinia cesatii*, *Ustilago duthiei*, etc. are found to be associated with the crop. It is also parasitized by *Striga lutea*. Ergot (*Claviceps* sp.) may present a significance in seed production.

4.20 HARVESTING

The crop is harvested two times a year, once in early to mid-summer and another in autumn. Physical harvesting is very much expensive.

4.21 YIELD AND SEED PRODUCTION

Marvel grasses yield about 15–60 q/ha dry matter; however under irrigated condition, the yield may go up to 170 q/ha. Green forage production ranges from 60 to 100 q/ha. The grass may produce 40–90 kg seed/ha.

4.22 NUTRITIVE VALUE

The grass contains 2.6–10.4% of crude protein, 34.9–45.5% crude fiber, and 7.1% lignin. Crude protein digestibility is 28–47% (Nooruddin et al., 1975). It also contains different minerals such as calcium 3.4 g, phosphorus 1.6 g, potassium 11.2 g, sodium 0.1 g, magnesium 1.1 g, manganese 46 mg, zinc 49 mg, copper 5 mg per kilogram of dry matter (Dougall and Bogdan, 1960; Sen, 1983). It is readily consumed by small and large ruminants (Coleman et al., 2004).

4.23 UTILIZATIONS

Marvel grass is very commonly used as fodder for ruminants such as sheep and goats. It can be used in pasture, cut, and carry system and suitable for hay and silage making. It produces good standing hay and can support seven sheep per hectare.

4.24 TOXICITY

There is no problem of toxicity.

4.25 COMPATIBILITY

It competes aggressively once it gets established and often suppresses other species.

KEYWORDS

- **marvel grass**
- **agronomic management**
- **yield**
- **quality**
- **utilization**

REFERENCES

Bogdan, A. V. *Tropical Pasture and Fodder Plants* (*Grasses and Legumes*); Longman: London and New York, 1977; pp 106–107.

Coleman, S. W.; Taliaferro, C. M.; Tyrl, R. J. *Old World Bluestems* (*Warm Season Grasses*); American Society of Agronomy: Madison, WI, 2004; pp 909–936.

Dougall, H. W.; Bogdan, A. W. The Chemical Composition of the Grass of Kenya-II. *E. Afr. Agric. For. J.* **1960,** *25* (4), 241–244.

Gill, R. S. *Personal Communication*. Punjab Agricultural University, Ludhiana, Department of Animal Science: Ludhiana, 1970.

Manidool, C. *Dichanthium annulatum* (Forssk.) Stapf. In *Plant Resources of South-East Asia No. 4: Forages*; Mannetje, L. 't., Jones, R. M., Eds.; Pudoc Scientific Publishers: Wageningen, the Netherlands, 1992; pp 181–183.

Mehra, K. L.; Celarier, R. P.; Harlan, J. R. Effects of Environment on Selected Morphological Characters in the *Dichanthium annulatum* Complex. *Proc. Oklah. Acad. Sci.* **1960,** *40*, 10–14.

Nooruddin; Roy, L. N.; Jha, G. D. Studies on the Digestibility and Nutritive Value of Marvel *Dichanthium annulatum*, Grass at the Flowering Stage. *Indian Vet. J.* **1975,** *52*, 350–352.

Sen, K. C. *The Nutritive Values of Indian Cattle Feeds and Feeding of Animals*; Indian Council of Agricultural Research: New Delhi, 1983; pp 1–30 (Bulletin No. 25).

CHAPTER 5

PANGOLA GRASS (DIGIT OR WOOLLY FINGER GRASS)

MD. HEDAYETULLAH[1*], PARVEEN ZAMAN[2], and RAJIB KUNDU[1]

[1]*Department of Agronomy, Bidhan Chandra Krishi Viswavidyalaya, Mohanpur, Kalyani 741235, Nadia, West Bengal, India*

[2]*Assistant Director of Agriculture, Pulse and Oilseed Research Sub-station, Department of Agriculture, Government of West Bengal, Beldanga 742133, Murshidabad, West Bengal, India*

[*]*Corresponding author. E-mail: heaye.bckv@gmail.com*

ABSTRACT

Pangola grasses are very much liked by all grazing animals and intake of the grass is satisfactory or good. The surplus herbage of the peak growth may be made into silage or hay but with difficulty. The pH value of the ensiled material is high and the seasonal high air humidity posses problems in hay making. Pangola grass is used extensively for pasture, hay, and silage. The grass is drought resistant and grows better with an annual rainfall between 1000 and 500 mm. The grass can grow on various types of soil from acid with 4.5 pH to alkaline with 8.5 pH but clayey soils are less suitable than loams. The average yield of green herbage is about 500–700 q/ha in the north and 800–900 q/ha in the south under improved package and practices. Maximum yield is up to 360 q of dry matter per hectare per year in six cuts.

5.1 BOTANICAL CLASSIFICATION AND BINOMIAL NAME

Kingdom: Plantae
Order: Poales

Family: Poaceae
Subfamily: Panicoideae
Genus: *Digitaria*
Species: *decumbens*
Binomial name: *Digitaria decumbens*

5.2 SYNONYMS

Digitaria eriantha subsp. *eriantha*, *Digitaria eriantha* subsp. *pentzii* (Stent) Kok, *Digitaria eriantha* subsp. *stolonifera* (Stapf) Kok, *Digitaria eriantha* var. *stolonifera* Stapf, *Digitaria geniculata* Stent, *Digitaria glauca* Stent, *Digitaria pentzii* Stent, *Digitaria pentzii* var. *minor* Stent, *Digitaria pentzii* var. *stolonifera* (Stapf) Henrard, *Digitaria polevansii* Stent, *Digitaria seriata* Stapf, *Digitaria setivalva* Stent, *Digitaria smutsii* Stent, *Digitaria stentiana* Henrard, *Digitaria valida* Stent, *Syntherisma eriantha* (Steud) Newbold.

5.3 COMMON NAME

Common finger grass, digit grass, woolly finger grass, smuts finger grass, giant *pangola grass*, and pangola grass.

5.4 INTRODUCTION

The name pangola grass has been derived from the Pangola River of Transvaal (South Africa). This grass was originally introduced to the United States in 1935; pangola grass has been given another name to another form of *Digitaria* from the same river area of Western Transvaal. The grass is popular because of the ease of establishments, good forage production, high nutritive value, and excellent palatability. The grass is drought resistant and grows better with an annual rainfall between 1000 and 500 mm. The grass can grow on various types of soil from acid with 4.5 pH to alkaline with 8.5 pH but clayey soils are less suitable than loams.

5.5 DESCRIPTION

Pangola grass (*Digitaria eriantha* Steud) is a tropical grass widespread in many humid tropical and subtropical regions, used extensively for grazing, hay, and silage. It is often considered to be one of the higher quality tropical grasses (Cook et al., 2005). *D. eriantha* is a monocot and in the family of Poaceae. It is perennial, sometimes stoloniferous. Each grass, erect or ascending, reaches the height between 35 and 180 cm. The lowest basal leaf sheaths are densely hairy or very rarely smooth.

5.6 GEOGRAPHICAL DISTRIBUTION

As an outstanding pasture grass, *Digitaria decumbens* soon spread to West Indies, Australia, West and East Africa, the Philippines, Hawaii, India, Pakistan, Malaysia, etc. in tropical, subtropical, and even temperate-warm countries.

5.7 CLIMATIC REQUIREMENTS

The grass is drought resistant and grows better with an annual rainfall between 1000 and 500 mm, decreases root growth at below 16°C and above 41°C soil temperatures. Low night temperatures adversely affect the growth and herbage yield of the grass, because during cold nights, starch grains accumulate in the leaf chloroplasts and hinder translocation of water soluble sugars. This starch accumulation occurs mainly in nontillering plants but in the actively tillering plants, the flow of the photosynthetic products from the leaves is not blocked even at 6–10°C. At high altitudes between 1200 and 500 m, the growth probably suffers because of cool nights.

5.8 SOIL REQUIREMENTS

The grass can grow on various types of soil from acid with 4.5 pH to alkaline with 8.5 pH but clayey soils are less suitable than loams. It withstands water logging only to a limited extent. During the cool season, especially at higher elevations, the productivity of pangola grass is very low, and a rotation cycle of 60 days or more is recommended (Fukumoto and Lee, 2003).

5.9 PLANT CHARACTERISTICS

The grass is a swallow-rooted perennial with long creeping stolons forming tufts. Its culms are simple or branched, 35–180-cm tall. The leaf blades are 5–60-cm long, 2–14-mm wide, glabrous, or hairy. *This grass* is sometimes stoloniferous or tufted and rhizomatous. The stolons spread over the surface of the ground and develop roots at the nodes. The stolons and stems are hairy. The inflorescence is a digitate or subdigitate panicle comprising 3–17 racemes, 5–20-cm long. The inflorescence is a terminal digitate panicle. Pangola grass is highly male and female is sterile. It is apomictic. Very few viable seeds are produced (Cook et al., 2005).

5.10 LAND PREPARATION

The grass does not like very deep and fine preparation of land. Three to four operations with a country plow, disk harrow, or cultivator are sufficient to get the tilth. Heavy soils may require more operations and rolling.

5.11 SOWING

Barring the period of minimum prevailing temperature below 16°C which retards growth, planting in the north, is usually during the monsoon and at many time of the year in the south and in the east. About 10–5-cm long mature stem cutting is broadcast on the prepared seed bed and mixed with the soil by trampling or plowing. The stems and rooted stolons may also be planted in furrows behind the plow or hoe to a depth of 5–10-cm deep. Planting, by rooted slips or soaked cutting (kept overnight under hessian cloth and protected from drying and direct sun), helps in easy and rapid establishment. Half the length of cutting should remain exposing over the soil surface after planting. In high rain fall areas, planting is done in raised beds, which are provided with free drainage, while in low rainfall areas, it is grown in low-lying situations, depression, or water drains (Chatterjee and Das, 1989).

5.12 SPACING

Spacing may vary from 25 cm × 25 cm to 50 cm × 50 cm or 50 cm × 100 cm depending upon the soil and rainfall. Between 10 and 20 q of fresh material is required for 1 ha of land.

5.13 CROP MIXTURES

High doses of N fertilizers increase pangola grass herbage and decrease or even eliminate the associated legumes (Whitney, 1970); several herbage legumes, for example, *Trifolium repens*, *Centrosema pubescens*, *Macroptilium atropurpureum*, *Desmodium intortum*, *Pueraria phaseoloides*, *Stylosanthes guianensis*, *Stylosanthes humulis*, *Calopogonium mucunoides*, *Teramnus labialis*, etc., showed reasonable success in association with pangola grass. This grass is also compatible with widely spaced subabool trees.

5.14 MANURING AND FERTILIZERS APPLICATION

Fertilizer N even up to 340 kg N/ha increases crude protein (CP) content and herbage yield of pangola grass. Fertilizers N decreases its tolerance to low temperatures and specially too frost. So, N should be applied with the advent of cool or cold season. Split application of N as foliar spray or side dressing after each grazing of cutting normally gives better result than single application. The herbage recovers 20–70% of the applied N, the highest recovery of N being at low-to-moderate rates.

P_2O_5 applied along gives little increase of herbage yield but it can be highly effective if applied along with high rates of N. Responses of K_2O are obtained when high rates of N are applied. K_2O increases tolerance top low temperatures unlike most of the other tropical grasses. Pangola grass response positively to Na which can replace a part of K_2O fertilizers.

Pangola grass is sensitive to Cu deficiency and response well to Cu oxide or $CuSO_4$. Application of P_2O_5 can increase C content in the plant. Sulfur can also improve the herbage yields and is generally applied in the form of gypsum. The grass is indifferent to liming and it tolerates Al in the soil. In India, 25 t of FYM per hectare may be applied every year either basal or with the break of the monsoon. At planting 30 kg N, 35 kg P_2O_5 and 30 kg K_2O may be applied. Then, 30 kg N/ha should be applied after each grazing or cutting. The above dose of P_2O_5 and K_2O should be repeated once in every year (Chatterjee and Das, 1989).

5.15 IRRIGATION AND DRAINAGE

Generally, pangola grass is a rainfed grass, but two to three irrigations are required during the summer to ensure to high yield of forage (Chatterjee and Das, 1989). The mean rainfall 900–1975 mm is optimum for its growth and development. In South Africa, it grows well under the 625–750-mm summer-dominant rainfall on moist, fertile, well-drained soils but is better suited to rainfalls of 1000–1200 mm in coastal regions. In high rainfall areas, planting is generally done in raised beds, which are provided with free drainage, while in low rainfall areas, it is grown in low-lying situations, depression, or water drains.

5.16 INTERCULTURAL OPERATION AND WEED MANAGEMENT

During the first 2–3 months of initial establishment, weeding and interculturing will be necessary for the grass to completely cover the ground. Later, the grass suppresses the weeds. To certain extent, pangola grass is sod bound or root bound. After 3 years, the pasture should be cultivated with a disk harrow or rotavator to renovate the grass and to kill weed. The same effect may be felt by burning the old stands immediately after the cold season.

5.17 HARVESTING

The grass may be harvested when it attained a height of about 40 cm; this normally takes 50–60 days. Being highly sensitive to low temperature, it gives four to five cutting in the north and six to seven cutting in the southern and eastern regions of the India.

5.18 GREEN FODDER YIELD

The average yield of green herbage is about 500–700 q/ha in the north and 800–900 q/ha in the south under improved package and practices.

Maximum yield is up to 360 q of dry matter (DM)/ha/year in six cuts. Herbage yield is influenced by season, fertilization, associated legumes, irrigation, frequency of cuts, and other management practices.

5.19 NUTRITIVE VALUE

CP content in DM of pangola grass ranges from 3% to 14%. Fertilizer N increases the content of CP but increase in interval between grazing or cutting decreases it. Early cuts in the season content more CP than subsequent cuts. Crude fiber (CF) content from 28% to 35% increases with the intervals of cuts. The herbage from longer intervals contains more stem and low leaf:stem ratio. The ether extract (EE) contents rage from 1% to 3% and nitrogen-free extract (NFE) from 41% to 63%. P content ranged from 0.11% to 0.16% depending on plant age and frequency of cutting, while Ca ranged from 0.50% to 0.76%. Total DM digestibility is high and varies from 47% to 67%, that of crude protein varies from 7% to 66%, of CF from 55% to 73%, of NFE from 41% to 71%, and of EE from 2% to 50%. Generally, pangola grass receiving moderate fertilization has a nutritive value sufficiently high for cattle and ship requirements. The vitamin A content in the forage is also adequate.

Animal production from the grass is usually high, average being 100–400 kg/ha with maximum and minimum records of 1275 and 175 kg/ha, respectively. The average milk production varies between 6 and 8 kg/ha/day from cows grazed at 2.57 head/ha. Average carrying capacity of the grass may be two to four cows of steers/ha and 6.5 steers on heavily fertilized pasture.

5.20 UTILIZATION

Pangola grasses are very much liked by all grazing animals and intake of the grass is satisfactory or good. The surplus herbage of the peak growth may be made into silage or hay but with difficulty. The pH value of the ensiled material is high and the seasonal high air humidity posses problems in hay making. Pangola grass is used extensively for pasture, hay, and silage. It withstands very heavy grazing (FAO, 2009). Regular grazing (2–3 week intervals) at 10–5 cm to 30–40 cm height is necessary to maintain the quality of pangola grass (Cook et al., 2005).

KEYWORDS

- **pangola grass**
- **agronomic package and practices**
- **fodder yield**
- **nutritive value**
- **utilization**

REFERENCES

Chatterjee, B. N.; Das, P. K. *Forage Crop Production Principles and Practices*; Oxford & IBH Publishing Co. Pvt. Ltd.: Kolkata, 1989.

Cook, B. G.; Pengelly, B. C.; Brown, S. D.; Donnelly, J. L.; Eagles, D. A.; Franco, M. A.; Hanson, J.; Mullen, B. F.; Partridge, I. J.; Peters, M.; Schultze-Kraft, R. *Tropical Forages*; CSIRO, DPI & F (Qld), CIAT and ILRI: Brisbane, Australia, 2005.

FAO. *Grassland Index. A Searchable Catalogue of Grass and Forage Legumes*; FAO: Rome, Italy, 2009.

Fukumoto, G. K.; Lee, C. N. Pangola Grass for Forage. Cooperative Extension Service, College of Tropical Agriculture and Human Resources, University of Hawaii at Manoa. *Livestock Manage*. **2003,** *4*, 1–4.

Whitney, A. S. Effects of Harvesting Interval, Height of Cut, Nitrogen Fertilization on the Performance of *Desmodium intortum* Mixture in Hawaii. Proceedings of 11th International Grasslands Congress, Surfers Paradise, Queensland, Australia, April 14–23, 1970; pp 632–636.

CHAPTER 6

CANARY GRASS (HARDING GRASS)

BULBUL AHMED[1*] and MD. HEDAYETULLAH[2]

[1]*Plant Physiology Division, Bangladesh Agricultural Research Institute (BARI), Joydebpur, Gazipur 1701, Bangladesh*

[2]*Department of Agronomy, Bidhan Chandra Krishi Viswavidyalaya, Mohanpur, Kalyani 741235, Nadia, West Bengal, India*

[*]*Corresponding author. E-mail: kbdahmed@gmail.com*

ABSTRACT

Canary grass is a perennial winter forage grass. Common names of canary grass are bulbous canary grass and Harding grass; it is a species of grass in the genus *Phalaris* of the Poaceae family. *Phalaris tuberose* is an important prospective forage crop in the hills of the temperate and subtemperate regions. Canary grass is a good pasture grass. Harding grass has more distinct rhizomes and an inflorescence that is compact at first but later becomes more open as the branches spread. Canary grass gives green herbage yield about 30–50 q/ha in first year while subsequent cuts may yield about 120 q/ha. It can give three to four cuttings during winter and spring. The succulent canary grass contains 3.5% crude protein and moisture content is about 80%. The grass has 12% digestible energy, 2% protein equivalent, and 9% starch equivalent.

6.1 BOTANICAL CLASSIFICATION

Kingdom: Plantae
Order: Poales
Family: Poaceae

Genus: *Phalaris*
Species: *tuberosa* (L.)
Binomial name: *Phalaris tuberosa* (L.)

6.2 SCIENTIFIC NAME

Phalaris tuberosa (L.).

6.3 COMMON NAME

Harding grass, bulbous canary grass, ribbon grass, gardener's garters, toowoomba canary grass, Kolea grass, etc.

6.4 INTRODUCTION

Canary grass is a perennial forage crop and a wild grass. Canary grass is primarily adapted for permanent hay or pasture on sites where it is too wet for good performance of other forage grasses. Common names of canary grass are bulbous canary grass and Harding grass; it is a species of grass in the genus *Phalaris* of the Poaceae family. *Phalaris tuberose* is an important prospective forage crop in the hills of the temperate and subtemperate regions. Canary grass is a good pasture grass. Harding grass has more distinct rhizomes and an inflorescence that is compact at first but later becomes more open as the branches spread. Canary grass gives green herbage yield about 30–50 q/ha in first year while subsequent cuts may yield about 120 q/ha. It can give 3–4 cuttings during winter and spring. The succulent canary grass contains 3.5% crude protein (CP) and moisture contains about 80%. The grass has 12% digestible energy, 2% protein equivalent, and 9% starch equivalent. The grass can be used as silage and hay. The grass gives satisfactory silage with molasses. Acetic acid fermentation is occurred but not lactic acid during fermentation. Good hay can be prepared from this grass. *Phalaris* species contain gramine, which can cause brain damage, other organ damage, central nervous system damage, and death in sheep.

6.5 ORIGIN AND GEOGRAPHICAL DISTRIBUTION

Originally a native of the Mediterranean regions, it is now grown commercially in several parts of the world for birdseed. This grass is an exotic grass that has been introduced from Australia and New Zealand. In India, it is propagated in Rajasthan, Himachal Pradesh, and the Nilgiris (Chatterjee and Das, 1989). It was introduced, and is now widely naturalized, into South Africa, Australia, New Zealand, Ireland, the United Kingdom, and the United States.

6.6 CLIMATIC REQUIREMENTS

Canary grass has wide adaptability of soil and climate. It prefers good rainfall but it thrives in low rainfall areas as well. The grass grows well in 20–30°C temperatures but it is also tolerant to drought, frost, and severe cold weather. It grows in hilly terrains where there is scarcity of good pasture grasses. In cold humid winter, it performs well compare to other grassy fodder crops. Although its initial growth is slow, once it established, it withstand heavy grazing. The exceptional qualities have been made by this grass in pasture land.

6.7 PLANT CHARACTERISTICS

The grass is perennial in nature. This large, coarse grass has erect, hairless stems, usually from 2 to 6-ft tall. The ligule is prominent and membranous, 0.6-cm long, and rounded at the apex. The gradually tapering leaf blades are 8.9–25.4-cm long, 0.6–1.9-cm wide, flat, and often harsh on both surfaces. The compact panicles are erect or sometimes slightly spreading and range from 7.6 to 40.6-cm long with branches 1.2–3.8-cm long. Single flowers occur in dense clusters in mid-April. Inflorescences are green or slightly purple at first and then become tan. The seeds are shiny brown.

6.8 LAND PREPARATION

The perennial grass requires thorough cultivation and land preparation which may need one deep plow with disk plow followed by three to four operations with a country plow or disk harrow. The land should be free from weeds.

6.9 SOWING

The direct seed sowing is generally done for its propagation. The seed germinates well near about 20°C. Sowing is, therefore, taken up in September and October. February and March sowing are also done when temperature fluctuates between 20°C and 30°C. The seeds are broadcast or sown in line.

6.10 SPACING AND DEPTH OF SOWING

The recommended spacing is 20 cm × 10 cm, but the row-to-row distance may be increased to 30 cm when the soil is poor and irrigation facility is not available. The tiny seed should not be placed below 2–2.5 cm soil surface.

6.11 SEED RATE

The seed rate varies from 2.5 to 4 kg/ha if it is a clean seed. A thousand seeds weigh about 1.13 g.

6.12 CROP MIXTURES

Canary grass belongs to the Poaceae family, since it is not suitable combination to grow with other grasses. A number of pasture legumes can grow with canary grass. It can be grown with lucerne, berseem, shftal, senji, metha, etc. in the winter and with cowpea, guar, rice bean, green gram, black gram, velvet bean, etc. in the summer. Legumes grown in mixtures with the grass seldom increase the yield of grass but they usually increase total dry matter and especially CP yields of the mixed herbage (Chatterjee and Das, 1989).

6.13 VARIETIES

The cultivated canary grass variety is common in growing areas such as AQ1, Uneta, and Australis.

6.14 MANURES AND FERTILIZERS

Phalaris tuberose responds well to nitrogenous fertilizers and produces over 30 kg dry matter or even 65 kg dry matter/kg of applied N. In natural grasslands, nitrogenous fertilizers respond well if *Phalaris tuberose* increases their herbage production. Because it grows best in at least moderately fertile soils, it is advisable to use an establishment application of 200–250 kg/ha of super phosphate and 60 kg/ha of muriate of potash on less fertile soils. Nitrogenous fertilizers like urea application as a basal at the rate of 100 kg/ha (46 kg/ha N) are beneficial in pure stands.

6.15 IRRIGATION AND DRAINAGE

The field should be irrigated when it required. The crop requires about 400–500 mm water per crop season. It requires 500 mm annual rainfall with good distribution from autumn to spring for optimal growth (Watson et al., 2000). However, it can grow in places where annual rainfall is as low as 300 mm provided the soil has good moisture-holding capacity (Dyer, 2005). For better crop establishment, presowing and postsowing irrigation are important. Subsequent irrigations should be given as and when necessary.

6.16 INTERCULTURAL OPERATION AND WEED MANAGEMENT

After first 2–3 months of crop establishment, the crop requires one or two weeding and interculture operation. Every year of the cool season, interculture operation is done by country plowing or hoe is necessary to control grassy, broad leaf, and sedge weeds and encourages the plant growth.

6.17 PEST AND DISEASES

Generally, no serious pest and diseases is occurred in *Setaria* grass. Attack by armyworm and pasture webworm can destroy much of the leaf, particularly young leaf. *Pyricularia* leaf spot is prevalent under hot humid conditions and can retard the growth of the plant.

6.18 HARVESTING

The young plants make slow initial growth in the first year of establishment. After that, they grow well and harvesting may be taken up at an intervals. The first cut can be taken after sowing or planting of 9–10 months. The subsequent cuts can be taken after every 5–8 weeks depending on the season and growth of the grass. Generally, seven to eight cuts can be taken per year from the same field best management practices. After each cutting, a stubble height of 8–10 cm is left for good regeneration.

6.19 GREEN FODDER YIELD

Canary grass gives green herbage yield about 30–50 q/ha in first year while subsequent cuts may yield about 120 q/ha. It can give three to four cuttings during winter and spring.

6.20 TOXICITIES

Phalaris species contain gramine, which can cause brain damage, other organ damage, central nervous system damage, and death in sheep. Leaves and seedlings contain the tryptamine and related compounds.

6.21 NUTRITIVE VALUE

The succulent canary grass contains 3.5% of CP and moisture content about 80%. The grass has 12% digestible energy, 2% protein equivalent, and 9% starch equivalent.

6.22 UTILIZATION

The grass can be used as silage and hay (Culvenor, 2007). The grass gives satisfactory silage with molasses. Acetic acid fermentation is occurred but not lactic acid during fermentation. Good hay can be prepared from this grass. In hills, green fodder fed to animals during winter months.

KEYWORDS

- **canary grass**
- **agronomic management**
- **fodder yield**
- **nutritive value**
- **toxicities**

REFERENCES

Chatterjee, B. N.; Das, P. K. *Forage Crop Production Principles and Practices*; Oxford & IBH Publishing Co. Pvt. Ltd.: Kolkata, 1989.

Dyer, D. Kolea Grass (*Phalaris aquatica* L.). *United State Department of Agriculture-Natural Resource Conservation Service, Plant Guide*, Plant Materials Center, Lockeford, CA and Reina O'Beck, California State Office, Davis, CA, 2005, pp 1–3.

Watson, R. W.; McDonald, W. J.; Bourke, C. A. Phalaris Pastures (Agfacts), 2nd ed., July, 2000. *Affect* **2000,** *2* (5), 1.

CHAPTER 7

ANJAN GRASS (AFRICAN FOXTAIL GRASS)

MD. HEDAYETULLAH[1*], B. BHATTACHARYA[2], and PARVEEN ZAMAN[3]

[1]*Department of Agronomy, Bidhan Chandra Krishi Viswavidyalaya, Mohanpur, Kalyani 741235, Nadia, West Bengal, India*

[2]*Seed Farm, AB Block, Bidhan Chandra Krishi Viswavidyalaya, Kalyani 741235, Nadia, West Bengal, India*

[3]*Assistant Director of Agriculture, Pulse and Oilseed Research Sub-station, Department of Agriculture, Government of West Bengal, Beldanga 742133, Murshidabad, West Bengal, India*

[*]*Corresponding author. E-mail: heaye.bckv@gmail.com*

ABSTRACT

Anjan grass is very nutritious and palatable. It is relished by all classes of livestock even at maturity. It is called buffel grass or African Foxtail and is a valuable tufted perennial grass in arid and semiarid areas characterized by severe drought, high temperature, low rainfall, and sandy soil. It is an excellent grazing perennial, suited to pasture and range lands. It is not very aggressive and is used for rehabilitating sand dunes and recuperating highly eroded areas. Its high soil-binding capacity is due to its clustered root system in the upper 8–10-cm layer of soil. The protein content ranges from 3% to 16%. The well calcium phosphorus ratio is 0.48–1.9%:0.41–1.05%. The crude protein percentage is 8.39, ether extract is 1.72%, crude fiber is 30.54%, total acid is 16.08, and nitrogen-free extract is 43.30%. Average green forage yield gives about 220–300 q/ha/year. The yield of pasture grass is about 150–200 q/ha/year. Oxalate levels can cause “big head" (*Osteodystrophia fibrosa*) in horses and oxalate poisoning in young or hungry sheep.

7.1 BOTANICAL CLASSIFICATION

Kingdom: Plantae
Order: Poales
Family: Poaceae
Subfamily: Panicoideae
Genus: *Cenchrus*
Species: *ciliaris* (L.)
Binomial name: *Cenchrus ciliaris* (L.)

7.2 COMMON NAME

Anjan (Hindi—anjan, dhaman baiba, charwa, kusa; Tamil—kolickattai pillu; Telegu—kusa), buffel grass, foxtail buffalo grass, blue buffalo grass (African), foxtail grass (English), bloubuffelgras (South Africa), or anjan grass (India).

7.3 BOTANICAL NAME

Cenchrus ciliaris (Linn.); *Synonymous*: *Pennisetum cenchroides* Rich and *Pennisetum celiare* Link

7.4 INTRODUCTION

It is called buffel grass or African Foxtail. It is a valuable tufted perennial grass in arid and semiarid areas characterized by severe drought, high temperature, low rainfall, and sandy soil. It is an excellent grazing perennial, suited to pasture and range lands. It is not very aggressive and is used for rehabilitating sand dunes and recuperating highly eroded areas. Its high soil-binding capacity is due to its clustered root system in the upper 8–10-cm layer of soil. The establishment is slow and the grass is very persistent. It survives extreme and prolonged drought but grows vigorously popular in moist areas where more productive grasses can be grown. The grass is very nutritious and palatable. It is relished by all classes of livestock even at maturity. Anjan grass is distributed in subhumid and semiarid tropics and subtropics and native to many countries in Africa, Asia, and the Middle East. Average green forage yield gives about 220–300 q/ha/year. The yields of pasture grass about 150–200 q/ha/year. The protein content ranges from 3% to 16%. The well calcium phosphorus ratio is 0.48–1.9%:0.41–1.05%. The crude protein percentage is 8.39, ether extract (EE) is 1.72%, crude fiber (CF) is 30.54%, total acid (TA) is 16.08, and nitrogen-free extract (NFE) is 43.30% (Chatterjee and Das, 1989). Oxalate levels can cause "big head" (*Osteodystrophia fibrosa*) in horses and oxalate poisoning in young or hungry sheep. Under good management conditions, the seed yield gives about 1–2 q/ha. The fodder seeds can be stored for 2–3 years.

7.5 PLANT CHARACTERISTICS

The grass is a perennial and tufted grass. Stems are erect or ascending up to 140-cm high, stout, or slender. Leaves are glabrous or hairy, green or bluish, flat, 7–30-cm long, and 2–5-cm wide. The inflorescence is a cylindrical raceme of loosely arranged spike 3–15-cm long and generally light brown in color. The spikes are surrounded by involucres of sterile spikelets modified in to bristles. Anjan grass is sometimes stoloniferous (Halvorson and Guertin, 2003), reproduced by seed sexually and apomictically. Seed set much more likely when plants cross-pollinate than when self-pollination occurs. Plants can propagate by rhizomes. Seeds are viable in the soil for up to 4 years. Seed dispersal occurs by water, wind, animals (livestock and wild), and humans (on clothing and via vehicles) (Halvorson and Guertin, 2003). Seed dormancy increases if water stress occurs when seeds are maturing. Seed dormancy decreases when soil fertility and temperature increase (Sharif-Zadeh and Murdoch, 2000).

7.6 GEOGRAPHICAL DISTRIBUTION

Cenchrus ciliaris is a native of India, Africa, Indonesia, and semiarid Europe (Chambers and Hawkins, 2002). The grass in the natural pastures and ranges lands of arid and semiarid regions of the tropical and subtropical

India and Africa, North Africa, Madagascar, Canary Islands, Arabia, Pakistan, Iran, Afghanistan, Indonesia, Queensland and South Wales of Australia, and to some extent in the USA. In India, it is found as a natural grass in the *Dichanthium/Cenchrus/Elyonurus* cover between 23°N and 32°N latitude and 60°E and 80°E longitude. It is distributed from sea level to 2000-m altitude.

7.7 CLIMATIC REQUIREMENT

Buffel grass can grow well in areas having 270–300 mm upward annual rainfall in a single wet season. It thrives all high temperatures up to 45–48°C, the optimum temperature for photosynthesis being about 35°C. The grass remains green even in the summer season with high temperature (48°C) but the stem becomes wiry. The grass is frost tolerant. Depending upon the cultivar, the duration of water logging adversely affects the field germination of seed, seedling emergence and crop performance. The grass can grow on alluvial, late rite, and red soil and medium black soils. It comes up best on calcareous red soils with pH about 7.5; it also does well in forest lands but cannot thrive on clayey soils or on late rite soils deficient in lime. It can withstand considerable soil moisture stress. Germination occurs at soil pH levels from 3.0 to 7.0 (Emmerich and Hardegree, 1996).

7.8 SOIL AND ITS PREPARATION

One deep plowing by mold board (MB) plow for deep tillage on its roots penetrates deeper in soil. MB plow also helps to turn over the soil; weeds and crop stubble and residue go deeper in soil. The land should be brought to fine tilth by repeated cross-harrowing or plowing. High salinity, freezing temperatures, heavy clay soil, deep sand, high water table, and poor surface drainage limit distribution of this grass (Hanselka, 1988).

7.9 VARIETIES

Different genotypes were tired at the CAZRI, Jodhpur; strain nos. 357 and 358 gave comparatively a higher yield of forage on light sandy soil with 250-mm rainfall. The strain numbers 226 and 362 were found better in high rainfall zone and in comparatively heavy soil. Varieties like Jodhpur local, IGFRI 3133, and IGFRI 3108 are most compatible grasses with legumes. Pusa giant is highly productive (600–650 q/ha) and more leafy with a high content of crude protein (14–15%). It is evolved from a cross between Rajasthan and American strain. It showed excellent recovery after grazing and cutting. It is also tolerant to cold. Pusa yellow is more persistent, aggressive, soft, and leafy under drier conditions than black anjan. Then, other promising strains are IGFRI-1, CAZRI-1, and Coimbatore-1. Some of the cultivars like Biloela, Molopo, Boorara, Lawes, Nunbank, Tarewinnber, and Higgins grass are tall, vigorous, stoloniferous, hard stemmed, and suitable for cattle grazing.

7.10 SOWING/PLANTING TIME

Under irrigated conditions, sowing can be done in any season and under rainfed condition, seed sowing may be done with premonsoon rain of May–June or with the break of monsoon. For transplanted crop, nurseries can be raised in early June and seedlings are ready for transplanting after 3–4 weeks.

7.11 SEED RATE AND SOWING METHOD

The rate of 5–6 kg/ha is maintained in sandy soils of hot and dry areas. The seed rate is increased to 7–10 kg/ha in loam to clay loam soils.

Anjan grass seed is fluffy and is therefore broadcast after mixing it with fine sand or moist soil for easy and uniform distribution. Sowing whole clusters of spikelets give better results. The seeding may also be done by *kera* method with a 1–2-cm depth. After seed placing, planking by wooden log is must for covering soil on seed for better germination. The seed should be soaked about 8 h before sowing for sowing erodible soils and gullies.

7.12 SPACING

Plant population depends on the soil type, fertility status, and rainfall. In light soil of low rainfall areas, plants are kept wide spacing of 75 cm × 75 cm. The spacing may be reduced to 60 cm × 60 cm in heavy to loamy soil with optimum moisture conditions. It may further reduced to 60 cm × 30 cm in high fertile soils and subhumid conditions. It is generally sown in *kera* method that the row-to-row spacing is maintained at 60–70 cm apart and the plants are thinned to 30–45 cm.

7.13 CROPPING SYSTEMS

The anjan grass can be grown in mixtures with relatively drought-resistant perennial or annual legumes like stylo, Lucerne, siratro, soja, Centro, lablab bean, moth, moong, cowpea, velvet bean, etc. The grasses like marvel grass, sewan grass, and Rhodes grass can also be grown together.

7.14 NUTRIENT MANAGEMENT (MANURES AND FERTILIZERS)

Adequate organic manure gives higher green forage production. Application of 10–15 t/ha cow dung manures or composts farmyard manure at land preparation. Application of chemical fertilizers at the rate of 30:25:30 kg $N:P_2O_5:K_2O$ at the time of final land preparation. After soil test, the dose of fertilizers may increase or decrease. Anjan grass responds well in application of phosphorus at the rate of 50 kg/ha.

7.15 WATER MANAGEMENT

Anjan grass is crop of dry areas and does not tolerate waterlogging conditions. Light irrigations during summer months show quick recovery and faster growth. Irrigation is also required in the postmonsoon season for good crop stand. During rainy season, land should be well drained.

7.16 INTERCULTURE OPERATION AND WEED MANAGEMENT

At early stage of crop, crop and weed competition is more common. One and two weeding at 25–30 days after sowing and second 45–50 days after sowing are very essential for good plant stands and weed free field. One or two hand weeding is beneficial to crop as soil stir up and stimulate the growth of the grass. Two weeding give more herbage yield compare to the one weeding.

7.17 INSECT-PEST AND DISEASE MANAGEMENT

The only major insect pest of buffel grass is the buffel grass seed caterpillar (*Mampava rhodoneura*), a paralid moth. It has been recorded in warmer, higher rainfall areas, feeding on the seeds, and webbing the heads together, but control is not considered economic on a broad acre basis. However, for seed crops, it can be controlled by spraying crops with methomyl 10 days after heads emerge. Buffel blight, caused by fungal pathogen *Pyricularia grisea*, and ergot (*Claviceps* spp.) affecting seed production, is the most important diseases of buffel grass (Perrott, 2000). Buffel blight causes "extensive losses" in monocultures, affecting the persistence of the grass through "ill-thrift," but its economic significance is not known. *Fusarium oxysporum* has also been found in association with buffel dieback (Makiela et al., 2003). Hall (2000) indicates that dieback from fungal attack occurs in isolated cases in Queensland so that the economic impact may not be great, although a small number of producers have reported production reduced to a third of previous levels in infected areas (Perrott, 2000).

7.18 HARVESTING

Anjan grass can be harvested 10–12 months after sowing. Height of cutting is maintained at 5–10 cm above the ground level for better yield and herbage quality. Frequent close cutting gives lesser yield of subsequent cut as well as carbohydrate content. Generally, first year of crops are not allowed grazing but light grazing is done in second year. Three to four cuts taken in second year and subsequently six cuts are taken every year. Plant is very tolerant of heavy grazing (Williams and Baruch, 2000), tolerates short periods of overgrazing; prolonged heavy grazing results in decreased root growth (Hanselka, 1988).

7.19 GREEN FODDER YIELD

The yield of green herbage is 90 q/ha/year when annual rainfall receives up to 300 mm; where annual rainfall ranges between 380 and 780 mm, green forage yield gives about 220–300 q/ha/year. The yields of pasture grass is about 150–200 q/ha/year.

7.20 SEED PRODUCTION

The seed yield depends on cultivar, rainfall pattern, and soil types. Matures are harvested by hand or by machine and separation of seed from spike is difficult. Generally, seeds are harvested in October and November. The seed crop is usually sown in wider space than fodder crop for good quality seeds. Under good management conditions, the seed yield gives about 1–2 q/ha. The fodder seeds can store 2–3 years.

7.21 NUTRITIVE VALUE

Anjan grass is relished by all classes of livestock. It is an excellent type maintenance quality fodder and can support milk yield without concentrate up to a limited extent. It contains 11% crude protein at young stage with suitable ratio of calcium and phosphorus. Neutral detergent fiber and acid detergent fiber content is 72.0% and 38.0%, respectively. It provides vary good hay since it retains its nutritive value even when ripe fully. Nutritive value of this grass is too high compared to others grasses. The protein content ranges from 3% to 16%. The well calcium phosphorus ratio is 0.48–1.9%:0.41–1.05%. The crude protein percentage is 8.39, EE is 1.72%, CF is 30.54%, TA is 16.08, and NFE is 43.30%. Anjan grass is considered a highly nutritious pasture grass in hot, arid regions. One hundred grams of green grass contains 110-g protein, 26-g fat, 732-g total carbohydrate, 319-g fiber, and 132-g ash. Also, 100 g of hay reportedly contains 74-g protein, 7-g fat, 792-g carbohydrate, 352-g fiber, and 17-g ash (Gohl, 1981).

7.22 UTILIZATION

This grass is widely used as cut fodder. It is well suited for grazing or ensiling. It makes very good hay after quick drying in the sun. It is mainly used as a permanent pasture but can be used for hay or silage. Not suited to short-term pasture because too difficult to remove and binds nutrient.

7.23 SPECIAL FEATURES (TOXICITIES)

There is some evidence that buffel grass can release allelopathic chemicals which alter soil properties to the extent that germination and growth of other plants are inhibited (Cheam, 1984). Oxalate levels can cause "big head" (*O. fibrosa*) in horses and oxalate poisoning in young or hungry sheep. However, with soluble oxalate levels of 1–2% in the DM, there is rarely a problem with mature ruminants.

KEYWORDS

- **anjan grass**
- **cultivation practices**
- **yield**
- **nutritive value**
- **utilization**

REFERENCES

Chambers, N.; Hawkins, T. O. *Invasive Plants of the Sonoran Desert: A Field Guides*; Sonoran Institute, Environmental Education Exchange, National Fish and Wildlife Foundation: Tucson, AZ, 2002; p 120.

Chatterjee, B. N.; Das, P. K. *Forage Crop Production Principles and Practices*; Oxford & IBH Publishing Co. Pvt. Ltd.: Kolkata, 1989.

Cheam, A. H. Allelopathy in Buffel (*Cenchrus ciliaris* L.). Part 1. Influence of Buffel Association on Calotrope (*Calotropis procera* (Ait.) W. T. Ait.). *Australian Weeds* **1984,** *3*, 133–136.

Emmerich, W. E.; Hardegree, S. P. Partial and Full Dehydration Impact on Germination of 4 Warm-Season Grasses. *J. Range Manage.* **1996,** *49* (4), 355–360.

Gohl, B. Tropical Feeds. Feed Information Summaries and Nutritive Values. *Animal Production and Health Series*; FAO: Rome, 1981; p 12.

Hall, T. J. History and Development of Buffel Grass Pasture Lands in Queensland. In *Proceedings of Buffel Grass Workshop*, Theodore, Qld, 21–23 February 2000; Cook, B., Ed.; Department of Primary Industries: Queensland, 2000; pp 2–12.

Halvorson, W. L.; Guertin, P. *U.S.G.S. Weeds in the West Project: Status of Introduced Plants in Southern Arizona Parks*, 2003.

Hanselka, C. W. Buffelgrass—South Texas Wonder Grass. *Rangelands* **1988,** *10* (6), 279–281.

Makiela, S.; Harrower, K. M.; Graham, T. W. G. Buffel Grass (*Cenchrus ciliaris*) Dieback in Central Queensland. *Agric. Sci.* **2003,** *16*, 34–36.

Perrott, R. In *Diseases of Buffel Grass*, Proceedings of Buffel Grass Workshop, Theodore, Qld, Feb 21–23, 2000; Cook, B., Ed.; Department of Primary Industries: Queensland, 2000; pp 19–20.

Sharif-Zadeh, F.; Murdoch, A. J. The Effects of Different Maturation Conditions on Seed Dormancy and Germination of *Cenchrus ciliaris*. *Seed Sci. Res.* **2000,** *10*, 447–457.

Williams, D. G.; Baruch, Z. African Grass Invasion in the Americas: Ecosystem Consequences and the Role of Ecophysiology. *Biol. Invasions* **2000,** *2* (2), 123–140.

CHAPTER 8

RHODES GRASS (ABYSSINIAN RHODES GRASS)

MD. HEDAYETULLAH* and SIBAJEE BANERJEE

Department of Agronomy, Bidhan Chandra Krishi Viswavidyalaya, Mohanpur, Kalyani 741235, Nadia, West Bengal, India

**Corresponding author. E-mail: heaye.bckv@gmail.com*

ABSTRACT

Rhodes grass is an important tropical grass widespread in tropical and subtropical countries. It is useful forage for pasture and hay, drought-resistant and very productive, and of high quality when young. It can grow in many types of habitat. The areas where overgrazing is a problem, it is the best choice for pasture, mainly for horse pasture as it has no oxalate problems. It is rarely affected by pests or diseases as well as can suppress the weed growth. "Katambora," a variety of Rhodes grass, is resistant to nematode. On an average, each cut gives about 100 q of green fodder/ha. Maximum yields are usually obtained in the second year of growth. Rhodes grass does not persist more than 5 years. Header-harvested yields of 100–200 kg/ha can be achieved from properly managed crops. Rhodes grass is primarily useful forage of moderate-to-high quality. It is grazed, cut for hay, or used as deferred feed but it is not suitable for silage.

8.1 BOTANICAL CLASSIFICATION

Kingdom: Plantae
Order: Poales
Family: Poaceae
Subfamily: Chloridoideae

Genus: *Chloris*
Species: *gayana* Kunth
Binomial name: *Chloris gayana* Kunth

8.2 BOTANICAL NAME

Chloris gayana Kunth; *Chloris abyssinica* Hochst. ex A. Rich.

8.3 COMMON NAME

Rhodes grass, Cecil Rhodes grass, Abyssinian Rhodes grass, Callide Rhodes grass, common Rhodes grass (English); chloris, Herbe de Rhodes (French); Capim de Rhodes (Portuguese); Grama de Rodas, pasto de Rodas, pasto Rhodes, Zacate gordura (Spanish); Rhodes grass (Afrikaans); Koro-Korosan (Philippines); banuko (Philippines)

8.4 INTRODUCTION

Rhodes grass is an important tropical grass widespread in tropical and subtropical countries. It is useful forage for pasture and hay, drought-resistant and very productive, of high quality when young. It can grow in many types of habitat. It is also cultivated in some areas as a palatable graze for livestock and a groundcover to reduce soil erosion and quickly revegetate denuded soil. It is tolerant of moderately saline and alkaline soils. Rhodes grass is one possible perennial improved grass which can be grown on farm permanent pasture or a short-to-medium-term pasture ley and used by smallholder farmers. It makes good hay if cut at or just before early flowering and provides better standover feed than buffel grass or the panics. It has better productivity and nutritive value when compared with natural pasture. It is also useful for erosion control by virtue of its spreading growth habit. It has wide adaptation capability. It thrives even in heavy drought-prone tracts and coastal areas also. It can be established easily with rapid reproduction through runners. The areas where overgrazing is a problem, it is the best choice for pasture, mainly for horse pasture as it has no oxalate problems. It is rarely affected by pests or diseases as well as can suppress the weed growth. "Katambora," a variety of Rhodes grass, is resistant to nematode.

8.5 PLANT CHARACTERISTICS

Rhodes grass is a perennial or annual tropical grass. It is a leafy grass, 1–2 m in height, highly variable in habit. Stems may be of two types; creeping stems which is 4–5 mm in diameter and upright stems with 2–4-mm diameter. Stem is smooth and shiny. The culms are tufted or creeping, erect or decumbent, sometimes rooting from the nodes. The roots are very deep, down to 4.5 m. The leaves are linear, with flat or folded glabrous blades, 12–50-cm long × 10–20-mm wide, tapering at the apex. The seed head has an open hand shape and encompasses 2–10 one-sided or double-sided racemes, 4–15-cm long. The inflorescences are light greenish brown in color and turn darker brown as they mature (Cook et al., 2005). The spikelets (over 32) are densely imbricated and have two awns. The fruit is a caryopsis, longitudinally grooved (Cook et al., 2005; Quattrocchi, 2006).

8.6 ORIGIN AND DISTRIBUTION

Rhodes grass is native to Africa but it can be found throughout the tropical and subtropical world (latitudinal range is between 18°N and 33°S). The name "Rhodes grass" was christened after Cecil Rhodes who was the pioneer of popularizing its cultivation in his estate near Cape Town in South Africa in 1895. It is now cultivated in tropical and some subtropical countries. This grass was introduced in 815 in India and thereafter it is grown in Maharashtra, Andhra Pradesh, Karnataka, and Madhya Pradesh.

8.7 CLIMATIC REQUIREMENTS

Rhodes grass grows well in the spring and summer and its rainfall requirement is 600–750 mm/year (Cook et al., 2005). This grass requires less rainfall and can survive in hot and arid places. The seed germinates quickly depending on temperature. Rhodes grass thrives well in places where annual temperatures range from 16.5°C to above 26°C, with maximum growth at 30°C/25°C (day/night temperature). Day lengths between 10 and 14 h are congenial for its growth and development. Optimal annual rainfall is about 600–750 mm with a summer-rainfall period (Cook et al., 2005). Rhodes grass can survive in areas where annual rainfall ranges between

310 and 4030 mm and where temperature extremes are 5°C and 50°C (Cook et al., 2005). Due to its deep roots, Rhodes grass can withstand long dry periods and up to 15 days of flooding (Cook et al., 2005; FAO, 2014).

8.8 SOIL AND ITS PREPARATION

Rhodes grass can grow in a variety of soil conditions. Some cultivars are tolerant of frost. Rhodes grass grows on a wide range of soils from poor sandy soils to heavy clayey alkaline and saline soils (more than 10 dS/m). This salt tolerance is particularly valuable in irrigated pastures where it can be cultivated without problem. Rhodes grass does better on fertile, well-structured soils and it prefers soil pH between 5.5 and 7.5. Establishment on acidic soils is difficult. It is tolerant of Li but not of Mn and Mg (Cook et al., 2005; FAO, 2014). Rhodes grass is a full sunlight species which does not grow well under shade (Cook et al., 2005; FAO, 2014). The ideal soil would be anything greater than a 4.3 pH level in terms of acidity and has a moderate aluminum tolerance. Less work is required to maintain this grass which means that the farmers can focus on other priorities. The land should be cleared from weeds and trees before plowing. The land should be plowed two to three times to get a fine and leveled seedbed. As the Rhodes grass seed is very small, it needs a well-prepared seedbed. Seedbed should be plowed and prepared well. Three to four plowings are enough to obtain fine and firm tilth for the grass.

8.9 VARIETIES

Some varieties of Rhodes grass are Pioneer, Katambora, Sanford, finecut, topcut, and Callide (for coastal area), etc. cultivated worldwide.

8.9.1 PIONEER

Pioneer also known as commercial Rhodes grass is an early flowering, erect plant with moderate leafiness. Because it will run to flower quickly throughout the growing season, its feed quality drops quickly. Pioneer has been superseded by Katambora.

8.9.2 KATAMBORA

Katambora is later flowering than Pioneer so remains more leafy and productive into autumn. It is also finer leaved and more stoloniferous.

8.9.3 CALLIDE

Callide is later flowering than Katambora, is less cold tolerant, and needs a higher rainfall than Pioneer or Katambora. Callide is more palatable and can be more productive than Pioneer or Katambora under conditions of higher fertility.

8.9.4 FINECUT

Finecut is a variety that has been selected for its improved grazing qualities. It has fine leaves and stems, is early flowering, of uniform maturity, and is high yielding. Finecut was derived from Katambora.

8.9.5 TOPCUT

Topcut, developed from Pioneer, has been selected for improved haymaking qualities. It has fine leaves and stems, is early flowering, of uniform maturity, and is high yielding.

8.9.6 CALLIDE

This variety is saline tolerant and grows well in coastal areas.

8.10 SOWING/PLANTING TIME

Under rainfed condition, sowing may be done with the premonsoon showers of June or with the break of monsoon. Seed germinates within 7 days after planting. Success of germination depends on depth of sowing. Sowing too deeply will result in failure or poor germination. In South

India, sowing can be done any time of the year, and in North India, it can commence by March (Chatterjee and Das, 1989).

8.11 SEED TREATMENT AND INOCULATION

The seed can be treated with fungicides like Thiram @2 g/kg of seed. The seed can be inoculated with azotobactor for nitrogen supply @20 g bacteria culture per kilogram seed. The bacteria culture can be inoculated with seed after 7 days of fungicide treatments. The bacteria culture can also be directly used in main field with farm yard manure for better bacteria growth.

8.12 SEED RATE, SPACING, AND SOWING METHOD

Rhodes grass is established vegetative (rooted slip) or from seed. The seed rate may be from 5 to 10 kg/ha. In case of drill sowing, the rows should be 60–75 cm apart and the plants may be spaced at 40–50 cm within rows for rooted cuttings. In case of vegetative propagation, about 28,000–30,000 rooted cuttings/ha are required for planting.

The seeds are used for propagation and vegetative part can also be used for propagation. The seeds may be broadcasted or drilled. The depth of sowing maintained is about 1.5–2 cm below the soil surface. The seeds are tiny and light, should be mixed with sand, moist soil, and saw dust for uniform distribution. Nurseries can also be raised to transplant about 6-week-old seedlings in the monsoon season.

Rooted slip is also used for its propagation. Rooted cuttings 15–30-cm long with two to three internodes can also be used for propagation (Chatterjee and Das, 1989).

In drill sowing, the rows may be 50–75 cm apart and the plants after final thinning may be spaced at 40–50 cm within the rows for rooted cutting. Hill to hill maintained is about 60 cm apart.

8.13 CROP MIXTURE

Rhodes grass can also be grown with legumes fodder like lucerne, cowpea, stylo, white clover, centro, etc. It can be grown with companion crops like maize, jower, bajra, teosinte, sudan grass, and cotton. Sometimes, it may also be intercropped with Subabul at early stage of Subabul. Oats, barley, and senji also intercropped with Rhodes grass in the cold season to protect the crop from frost (Chatterjee and Das, 1989). In Australia, it has been mixed with butterfly pea (*Clitoria ternatea*) for revegetation purpose (Cook et al., 2005).

8.14 NUTRIENT MANAGEMENT (MANURES AND FERTILIZERS)

Farmyard manure @10–15 t/ha is generally applied at the time of sowing. A balanced application of 40–50 kg N, 25 kg P, and 25 kg K/ha is recommended to enhance the nutritive value and yield. Then, 25–30 kg N/ha is applied after each cutting. The Rhodes grass is responded well in cattle shed wash with irrigation water.

8.15 WATER MANAGEMENT

It is often grown under irrigation in arid and semiarid areas. At establishment, the crop requires irrigation with 2 weeks interval in summer season and 3–4 weeks interval in the postmonsoon season.

8.16 WEED MANAGEMENT

At initial stages, weed competition is more. So, removing weeds at early stage of Rhodes grass production is very crucial. One or two weedings are important to reduce weed competition till the grass grows vigorously to cover the field. Application of 2,4-D at postplanting stage is also effective and controls the young broad-leaved weeds. Rhodes grass spreads readily in rainforest fringes in Queensland (Australia), where it produces seeds profusely and develops so quickly that it smothers native species and forms almost pure stands. However, *Rhodes grass* was shown to suppress summer weeds and has been considered helpful for controlling their development (Moore, 2006).

8.17 INSECT-PEST AND DISEASE MANAGEMENT

Rhodes grasses are severely attacked by armyworm and pasture webworm can destroy much of the leaf, particularly young leaf—largely restricted to coastal areas. It is rarely affected by pests or diseases as well as can suppress the weed growth. "Katambora," a variety of Rhodes grass, is resistant to nematode. Generally, this grass is not affected by serious diseases.

8.18 HARVESTING

To obtain high quantity and quality feed, the Rhodes grass should be harvested generally at 50% flowering stage. Rhodes grass can be harvested 4–6 months after planting. In North India, 4–5 cuts, and, in the South India, 8–10 cuts are done.

8.19 GREEN FODDER YIELD

On an average, each cut gives about 100 q of green fodder/ha. Maximum yields are usually obtained in the second year of growth. Rhodes grass does not persist more than 5 years (Chatterjee and Das, 1989). Based on several studies, the yields of Rhodes grass generally range from 100 to 120 q/ha/year, depending on variety, soil fertility, environmental conditions, and cutting frequency.

8.20 SEED PRODUCTION

Seed crops of Rhodes grass can be either direct headed or swathed. Harvesting should begin when seed starts to drop out of the tips of the seed heads. Up to three crops per year can be produced in most cultivars. Rhodes grasses are fertilized with 50 kg/ha N on fertile soils or 100–150 kg/ha N on infertile soils. Header-harvested yields of 100–200 kg/ha can be achieved from properly managed crops.

8.21 NUTRITIVE VALUE

Crude protein content of the Rhodes grass ranges between 3% and 17% depending on harvesting time and fertilization. Crude fiber content usually varies 25–30% depending on cutting stages and in late cut it may be high as 45% (Chatterjee and Das, 1989).

8.22 UTILIZATION

Rhodes grass is primarily useful forage of moderate-to-high quality. It is grazed, cut for hay, or used as deferred feed but it is not suitable for silage. It can form pure stands or is sown with other grasses or legumes. Many cultivars have been developed to suit different cultivation conditions or end uses, such as early, late, and very late flowering cultivars. Prostrate cultivars are suitable for grazing and erect cultivars are adapted to hay (Cook et al., 2005; FAO, 2014). *Chloris gayana* is useful as a cover crop and soil improver, as it improves fertility and soil structure and helps to decrease nematode numbers (Cook et al., 2005).

KEYWORDS

- **Rhodes grass**
- **agronomic package and practices**
- **fodder yield**
- **nutritive value**
- **utilization**

REFERENCES

Chatterjee, B. N.; Das, P. K. *Forage Crop Production Principles and Practices*; Oxford & IBH Publishing Co. Pvt. Ltd.: Kolkata, 1989.

Cook, B. G.; Pengelly, B. C.; Brown, S. D.; Donnelly, J. L.; Eagles, D. A.; Franco, M. A.; Hanson, J.; Mullen, B. F.; Partridge, I. J.; Peters, M.; Schultze-Kraft, R. *Tropical Forages*; CSIRO, DPI & F (Qld), CIAT and ILRI: Brisbane, Australia, 2005.

FAO. *Grassland Index. A Searchable Catalogue of Grass and Forage Legumes*; FAO: Rome, Italy, 2014.

Moore, G. *Rhodes Grass*. Department of Agriculture and Food, Western Australia. Bulletin 4690: Perth, 2006.

Quattrocchi, U. *CRC World Dictionary of Grasses: Common Names, Scientific Names, Eponyms, Synonyms, and Etymology*; CRC Press, Taylor and Francis Group: Boca Raton, FL, 2006.

PART II
Leguminous Forages

CHAPTER 9

GRASS PEA (INDIAN VETCH)

RAGHUNATH SADHUKHAN[1*], MD. HEDAYETULLAH[2], and PARVEEN ZAMAN[3]

[1]*Department of Genetics and Plant breeding, Bidhan Chandra Krishi Viswavidyalaya, Mohanpur 741252, West Bengal, India*

[2]*Department of Agronomy, Bidhan Chandra Krishi Viswavidyalaya, Mohanpur, Kalyani 741235, Nadia, West Bengal, India*

[3]*Assistant Director of Agriculture, Pulse and Oilseed Research Sub-station, Department of Agriculture, Government of West Bengal, Beldanga 742133, Murshidabad, West Bengal, India*

[*]*Corresponding author. E-mail: drsadhukhan@gmail.com*

ABSTRACT

Lathyrus (*Lathyrus sativus* L.), the grass pea or chickling pea, is also called *khesari* and *teora* in Hindi, *kasari* in Bengali, and Kisara in Nepali. The grass pea is widely adapted in arid areas and contains high levels of protein. Rain-fed crop gives green fodder yield potentiality of about 200–400 q/ha. The main limitation of grass pea is the neurotoxin ODAP. Lathyrus has multiple usages. It is used as a broken pulse (dal), besan (floor), animal feed, green fodder, and green leafy vegetables. Honeybees also visit to Lathyrus field to collect nectar.

9.1 BOTANICAL CLASSIFICATION

Kingdom: Plantae
Orders: Fabales
Family: Fabaceae

Subfamily: Faboideae
Tribe: Vicieae
Genus: *Lathyrus*
Species: *sativus*
Binomial name: *Lathyrus sativus* L.

9.2 COMMON NAME

Indian Vetch, Lathyrus, Chickling Pea, Grass Pea

9.3 BOTANICAL NAME

Lathyrus sativus L.

9.4 INTRODUCTION

Lathyrus (*Lathyrus sativus* L.), the grass pea or chickling pea, is also called *khesari* and *teora* in Hindi, *kasari* in Bengali, and *Kisara* in Nepali. It is traditionally used as food, feed, and fodder crop belonging to the family Fabaceae, subfamily Papilionoideae, tribe Vicieae. It is commonly known as *khesrari*, *lakhadi*, or *tewra.* The genus *Lathyrus* is large with 187 species and subspecies being recognized. Among these, four species, namely, *L. sativus*, *Lathyrus odoratus*, *Lathyrus ochyrous*, and *Lathyrus aphaca*, are found in India. However, only one species *L. sativus* is widely cultivated as a food crop, while other species are cultivated to a lesser extent for both food and forage. Several species are regarded as ornamental plants, especially the sweet pea (*L. odoratus*). The grass pea is endowed with many properties that combine to make it an attractive food crop in drought-stricken, rain-fed areas where soil quality is poor and extreme environmental conditions prevail. Despite its tolerance to drought, it is not affected by excessive rainfall and can be grown on land subject to flooding and that's why it is predominantly grown as a relay crop, popularly known as *utera*, in rice field. It has a very hardy and penetrating root system and therefore can be grown on a wide range of soil types, including very poor soil and heavy clays. This hardiness, together with its ability to fix atmospheric nitrogen, makes the crop one that seems designed to grow under adverse conditions. Because of the above unique characteristics of the crop, it is popular among the farmers and attracted them to grow this crop despite official discouragement (Chatterjee and Das, 1989).

The grass pea is widely adapted in arid areas and contains high levels of protein. The grass pea is important among forage crops which are grown in temperate countries and at high altitudes as in Jammu and Kashmir. Indian vetches are usually grown as *paira or utera* crop or relay crop with no tillage in wet rice fields. At global level, Lathyrus is a minor pulse crop; however, in some of the countries like India, Bangladesh, and Nepal, it is an important food and fodder crop. In South Asian countries, grass pea is commonly grown for both grain and fodder purposes. However, the crop has gained more importance for use as animal feed than for use as human food. Animal feed from grass pea is usually composed of ground or split grain or flour and is used primarily to feed lactating cattle or other draft animals. Human diets include Lathyrus as grains that are boiled and then either consumed whole or processed for split *dal* (Ahlawat et al., 2000).

9.5 PLANT CHARACTERISTICS

Grass pea is a much-branched, straggling or climbing, herbaceous annual, with a well-developed taproot system, the rootlets of which are covered with small, cylindrical, branched nodules, usually clustered together in dense groups. The stems are slender, 25–60-cm long, quadrangular with winged margins. Stipules are prominent, narrowly triangular to ovate with a basal appendage. The pinnate leaves are opposite, consisting of one or two pairs of linear-lanceolate leaflets, 5–7.5 cm × 1 cm, and a simple or much-branched tendril. Leaflets are entire, sessile, cuneate at the base, and acuminate at the top. The flowers are axillary, solitary, about 1.5-cm long and may be bright blue, reddish purple, red, pink, or white. The peduncle is 3.0–5.0-cm long with 2 min bracts. Flowers have a short and slender pedicel. Calyx teeth are longer and glabrous. Tube is 3-mm long with five lobes, subequal, and triangular. Standard petal is erect and spreading, ovate 15 mm × 18 mm, finely pubescent at upper margin, clawed. Wings are ovate, 14 mm × 8 mm, clawed, and obtuse at top. Keel is slightly twisted, boat-shaped, 10 mm × 7 mm, entirely split dorsally, ventrally split near the base. Color is lighter shaded than wing and standard. Stamens are diadelphous (9 + 1) with vexillary stamens free, 9-mm long, winged at base, apical part filiform, slightly winged. Staminal sheath is 6-mm long, with free filaments of uniform length. Anthers are elliptoid, 0.5-mm long, and yellow. Ovary is sessile, thin, 6-mm long, pubescent with 5–8 ovules. Style is abruptly upturned, 6–7-mm long, widening at tip, and bearded below the stigma. Stigma is terminal, glandular-papillate, and spatulate.

Pods are oblong, flat, slightly bulging over the seeds, about 2.5–4.5 cm in length, 0.6–1.0 cm in width and slightly curved. The dorsal part of the pod is two winged, shortly beaked, and contains three to five small seeds. Seeds are 4–7 mm in diameter, angled, and wedge-shaped. Color is white, brownish-gray, or yellow, although spotted or mottled forms also exist. Hilum is elliptic and cotyledons are yellow to pinkish yellow. The germination is hypogeal, the epicotyl purplish-green. The first two leaves are simple. The first leaf is small, scale-like, often fused with two lateral stipulae. The second leaf is sublate, connected at the base with stipulae. Normally, there are five seeds per pod. Seeds are wedge-shaped and angular.

9.6 ORIGIN AND DISTRIBUTION

There were two separate centers of origin of this crop. One was the Central Asiatic Center which includes northwest India, Afghanistan, the Republics of Tajikistan and Uzbekistan, and western Tian-Shan. The second was the Abyssinian Center. However, its domestication began in the Balkan Peninsula as a consequence of the Near East agriculture expansion into the region and its cultivation has now spread to include marginal lands of the Central, South, and Eastern Europe, in Crete, Rhodes, Cyprus, and in West Asia and North Africa (Syria, Lebanon, Palestine, Egypt, Iraq, Afghanistan, Morocco, and Algeria). The species are found both in the Old World and the New World. The center of diversity for Old World species in the genus are in Asia Minor and the Mediterranean region. According to De Candolle, *Lathyrus* is endemic to the region from Caucasus to north of India. It has been reported by many authors that the origin of *Lathyrus* was unknown as it was thought that the natural distribution had been completely obscured by cultivation, even in southwest and central Asia, its presumed center of origin. However, it is now suggested that the crop originated in the Balkan Peninsula. There are reports of wild *L. sativus* in Iraq but it is not clear if these are indeed wild or escapes from cultivation. From archeological evidence, it is suggested that the domestication of

Lathyrus dates to the late Neolithic and precisely to the Bronze Age. Prior to domestication, the crop was presumably present as a weed among other pulse crops (Ali and Mishra, 2000).

9.7 CLIMATIC REQUIREMENT

The crop is easily grown in both high and low rainfall areas. Thus, it grows well in Indian subcontinent including India, Bangladesh, Pakistan, and Nepal. It is also cultivated in China and in many countries of Europe, the Middle East, and Northern Africa. Grass pea, inherently capable of withstanding temperature extremes, is grown across diverse regions that receive an average annual precipitation ranging from 300 to 1500 mm. In addition to remarkable tolerance to drought, Lathyrus has tolerance to excess precipitation and flooding. It has a hardy and penetrating root system suited to a wide range of soil types including very poor soil and heavy clays (Ali and Mishra, 2000). It has inherent capacity of withstanding temperature extremes and is grown across diverse regions. In addition to remarkable tolerance to drought, it has tolerance to excess precipitation and flooding.

9.8 SOIL AND ITS PREPARATION

The crop grows well in wide range of soils except very acidic soil. Clayey soil that remains wet for a long time suits the crop well. The crop can withstand excess soil moisture, even standing water in the rice field at the time of sowing. Lathyrus is highly drought-tolerant crop and can withstand adverse soil water conditions.

Lathyrus can grow in minimum field preparation. Under good field preparation, Lathyrus gives higher yield compare to the unprepared field. The field should require one or two plowings followed by cross harrowing and planking.

9.9 VARIETIES

The traditional local land races contain higher concentration of ODAP (β-*N*-oxalyl-L-α,β-diaminopropionic acid). Therefore, the major thrust in varietal development program was laid on high yielding genotype with lower ODAP (below 0.1%) content. Some varieties of grass pea are Pusa 24, LSD 3, BIO L-212, Prateek, Mahateora, and Pusa 24.

9.10 SOWING TIME

The crop is generally sown in October and November and harvested in March. Lathyrus is a winter season crop adapted to the subtropics or temperate climates. December sowing crops are decreasing its yield. In many cases, this produces excessive biomass growth as the plant continues vegetative growth, usually until freezing terminates growth. Grass pea is usually grown after a crop of rice in South Asia and is broadcast into the standing rice crop approximately 15 days before harvest. The grass pea then germinates and grows on the residual soil moisture. This demonstrates a very hardy root system that allows the crop to grow under harsher conditions than do many other pulse crops.

9.11 SEED RATE AND SOWING METHOD

Grass pea in India, Bangladesh, Nepal, and Pakistan is generally sown at a rate of 40 kg/ha under "utera" or paira conditions. There is no land preparation as the seeds are broadcast by hand into standing water in rice fields before the rice is harvested. When it is sown on rainfed land, the soil is usually prepared by two plowings. The seeds are broadcast and planked about 2 weeks after the first plowing. The production of grass pea in Southeast Asia has often been observed in marginal areas that are often waterlogged, lowland, or rice-growing areas and plays an important role in increasing the cropping intensity. In such areas, farmers cannot produce a good winter crop of wheat, oilseeds, or other winter legumes. It is mostly sown into almost mature rice fields during November when field moisture is at a totally saturated condition. Usually, the seeds are soaked overnight and mixed with the fresh cow dung before sowing (broadcasting). Growers appear to be under the impression that mixing cow dung with the seeds protects the seeds from birds and insects and also enhances germination (Chatterjee and Das, 1989).

9.12 SEED TREATMENT AND INOCULATION

Seed should be treating with fungicides captan or Thiram @3 g/kg seeds is done 5–6 days before *Rhizobium* inoculation. Seed should be inoculated before sowing with *Rhizobium* sp. (cowpea miscellaneous group) to nodulate *Lathyrus* for fixation of atmospheric nitrogen. Inoculation of seeds is done about 12 h before sowing.

9.13 CROPPING SYSTEMS

Lathyrus is seldom grown pure except under *utera* cultivation in rice fields. It is grown as a mixed crop or intercrop with rainfed wheat, gram, barley, or linseed in dry uplands and the seeds are generally sown broadcast. Crop mixture is an important agronomic practice for getting assurance of produce under moisture stressed condition. Mustard + grass pea are good crop mixture. In rain-fed conditions, Lathyrus is along with barley or chick pea or rice—*khesary* rotation system. Generally, after harvesting of rice, grass pea is followed. After harvesting of jute, grass pea is grown under residual soil moisture (Prasad, 2006).

9.14 NUTRIENT MANAGEMENT (MANURES AND FERTILIZERS)

Lathyrus is neglected and minor crops in India and abroad too but its food, fodder, and vegetables value is too high. Under least management practice, this crop gives yield higher compare to others pulse crops. The farmers have not been much interested to use manures and fertilizers for this crop. It will give good results under ideal package and practices. Lathyrus gives better response to the application of nitrogen and phosphorus under relay cropping and an application of 10 kg N and 30 kg P_2O_5/ha.

9.15 WATER MANAGEMENT

Lathyrus can withstand in water stress conditions. Lathyrus is grown in dry land as well as lowland condition under residual soil moisture or in winter rains. In low land situation, Lathyrus is sown in standing paddy crop 15 days before harvest with 3–5 cm deep water. Also, one irrigation at vegetative stage is recommended if crops show water stress.

9.16 WEED MANAGEMENT

Crop weed competition is more common in early stage of crop growth (25–35 days after sowing). Lathyrus is a good crop and can cover or suppress the weed significantly. Normally in most growing areas of Southeast Asia, weeds are controlled by hand weeding. This crop often is considered a low input crop with lower returns and therefore weeding may not take place. Lathyrus was not competitive with weeds, especially when moisture was a limiting factor to plant growth. Herbicides which might safely control annual weed in *Lathyrus* have been identified. Pre-emergence herbicides like pendimethalin may be applied @ 0.75–1.0 kg a.i./ha.

9.17 DISEASE AND INSECT-PEST MANAGEMENT

Aphids (e.g., *Aphis craccivora*) are reported to be a major pest in India, Bangladesh, and Ethiopia. Powdery mildew (*Erysiphe polygoni* DC) and downy mildew (*Peronospora lathyri-palustris* Gaumann) are the two major diseases that infect grass pea. Powdery mildew (*Erysiphe pisi*) is an important disease of grass pea in India, causing economic losses of grain yield. This region has very congenial climatic conditions for endemic outbreak of this pathogen. Downy mildew is reported as a serious disease in Southeast Asia. The severity and incidence of root rot and wilt diseases (*Fusarium* spp.) ranged between 5–9% and 52.5%, respectively. Azadirection 1% EC @ 4–5 mL/L of water is very effective for controlling insects at an interval of 7–10 days. For controlling diseases, seed treatment is the effective way.

9.18 HARVESTING

The leaves turn yellow and the pods turn gray when mature. Pod splitting, which leads to premature shattering of seeds, is common in small types found on the Indian subcontinent. There, the plants are pulled out by hand or cut with a sickle near the base. The plants are then stacked and allowed to dry in the field or on the threshing floor for 7–8 days. The plants are spread out on the threshing floor and beaten with sticks. The seeds are more resistant to damage during harvesting operations than are field peas. It is a common practice to use cattle to help thresh the pods by trampling under the feet. The seed is then winnowed and cleaned. The seeds may be dried for several days before being stored. The straw and chaff are used for cattle feed, generally mixed with rice straw. The growers value the straw as a feed stuff as it is an important source of protein. In many cases, the value of the straw can be as great as that of the grain and thus is a major consideration for the farmer in the production of this pulse (Chatterjee and Das, 1989).

9.19 YIELD

Although the small-seeded types are desired for "utera" production, it can be seen that they are not necessarily higher yielding under these conditions. The amount of seed that is used per area of production does become important as the small-seeded types need significantly less than do large-seeded types to produce the same stand. As a rainfed crop, it shows green forage yield potentiality of 200–400 q/ha.

9.20 NUTRITIVE VALUE

Grass pea is nutritionally wealthy having 7.5–8.2% water, 48.0–52.3% starch, 25.6–31.4% protein, 4.3–7.3% acid detergent fiber, 2.9–4.6% ash, 0.58–0.80% fat, 0.07–0.12 mg/kg calcium, 0.37–0.49 mg/kg phosphorus, 18.4–20.4 lysine mg/kg, 10.2–11.5 mg/kg threonine, 2.5–2.8 methionine mg/kg, and 3.8–4.3 mg/kg cysteine.

However, its valuable nutritional aspect is rendered by the presence of neurotoxin β-ODAP (β-*N*-oxalyl-α,β-diaminopropionic acid) which make its grain unsuitable for human consumption if present in higher quantity. The neurotoxin β-ODAP is commonly present in seeds of all examined grass pea genotypes and responsible for Lathyrism, an irreversible paralyzing disease. This is the major limiting factor in the nutritional up-gradation and production of this valuable crop. The safe content of β-ODAP for human consumption is lower than 0.2% and only few varieties like

Ratan, Prateek, and Mahateora in India, and BARI Khesari-1 and BARI Khesari-2 in Bangladesh are available with <0.10% β-ODAP content. Furthermore, there appear to be fairly high levels of trypsin inhibitor in grass pea compared to many of the food legumes, the notable exception being soybeans which is another limiting factor of this crop. The presence of antinutritional factor like trypsin and chemotrypsinin in legume seeds decreases the digestibility of protein and causes pancreatic hypertrophy.

9.21 UTILIZATION

Lathyrus has multiple usages. It is used as a broken pulse (*dal*), besan (floor), animal feed, green fodder, and green leafy vegetables. Honeybees also visit to Lathyrus field to collect nectar.

9.22 TOXICITIES

The main limitation of grass pea is the neurotoxin β-*N*-oxalyl-α-β-diaminopropionic acid (β-ODAP). If the seeds are consumed as a major part of the diet for an extended period, irreversible crippling can occur. This remained a major limiting factor in the production of this valuable crop until such time as low- or zero-toxin lines were released. Very few sources with ODAP levels at 0.01% or lower are available which a major limitation for most breeding programs is. Low ODAP varieties released in India like Nirmal, Bio L-212 (Ratan), Pusa-24, Mahatura, and Pratick contain a neurotoxic nonprotein amino acid called β-ODAP(β-*N*-oxalyl-α-β-diaminopropionic acid) which causes neurolathyrism in humans and animals that causes an irreversible paralysis of the lower limbs in human and the four limbs in animal and is known as Lathyrism. The disease was recorded first in the Narwhal area of district Sialkot. The content of neurotoxin in the seed of local varieties can be anywhere between 0.37% and 1.2%. The other forms of the toxic substances in *L. sativus* are β-*N*-oxalyl amino-L-alanine (BOAA). Neurotoxin concentration is lower in *L. cicera* than in *L. sativus*. In *L. cicera*, it ranges from 0.04% to 0.76%. These values are genotype-dependent and show a little environment interaction. The safe content of ODAP for human consumption is lower than 0.2%. The most commonly used method for determining ODAP was the spectrophotometric method developed by Rao; for this reason, the method is also called *Rao-method* proposed by Prof. Lambein. The method involved the alkaline hydrolysis of ODAP to yield α,β-diaminopropionic acid, which in turn is complexed with *o*-phthalaldehyde in the presence of ethanol to form a colored product that is quantified at 420 nm. However, the method does not differentiate between the two isomers of ODAP (α-ODAP and β-ODAP). α-ODAP is a nontoxic or much less toxic isomer and easily formed from β-ODAP during heating. However, whenever new methods are developed for analysis of ODAP, the standard for comparison is nearly always the "Rao-method."

Lathyrus product has been banned in many countries due to the neurotoxin presence. However, due to the importance of these crops in developing countries, these countries have established breeding programs mainly focused on getting a genotype with high seed yield and low toxicity. There are two ways to have a genotype with high seed yield and low toxicity, first is using the genetic engineering to produce transgenic plants and the second way is through eliminating the toxic substance by careful selection so far and through hybridization between low and high toxin varieties.

9.23 FUTURE PROSPECTS OF LATHYRUS

Grass pea has a number of unique features that make it attractive to farmers and the consumers.

These are as follows:

- Adapted under harsh environment conditions such as excess moisture and drought.
- A high level of protein which normally ranges from 26% to 32%.
- Good taste which is used in snacks, various food products as well as component of the regular diet as *dal*.
- Minimizing or removal of neurotoxin content from both vegetative and reproductive parts will render the crop a more attractive fodder and food.
- A high biological nitrogen fixation ability which allows the crop to be an important component in sustainable cropping system.
- Used as forage or fodder for animals, both as a primary crop and as the residue after threshing.

- The grains are used as human food or as animal feed.
- Requires very low inputs for its cultivation. Virtually, seed is the only investment.

FIGURE 9.1 **(See color insert.)** Lathyrus.

KEYWORDS

- **grass pea**
- **agronomic package and practices**
- **fodder yield**
- **nutritive value**
- **toxicities**

REFERENCES

Ahlawat, I. P. S.; Prakash, O.; Saini, G. S. *Scientific Crop Production in India*; Aman Publishing House: Meerut, India, 2000.
Ali, M.; Mishra, J. P. *Technology for Production of Winter Pulses*; IIPR: Kanpur, 2000; p 62.
Chatterjee, B. N.; Das, P. K. *Forage Crop Production Principles and Practices*; Oxford & IBH Publishing Co. Pvt. Ltd.: Kolkata, 1989.
Prasad, R. *Text Book of Field Crop Production*; ICAR: New Delhi, 2006.

CHAPTER 10

MOTH BEAN (DEW BEAN)

PARVEEN ZAMAN[1*] and MD. HEDAYETULLAH[2]

[1]*Assistant Director of Agriculture, Pulse and Oilseed Research Sub-Station, Department of Agriculture, Government of West Bengal, Beldanga 742133, Murshidabad, West Bengal, India*

[2]*Department of Agronomy, Bidhan Chandra Krishi Viswavidyalaya, Mohanpur, Kalyani 741235, Nadia, West Bengal, India*

[*]*Corresponding author. E-mail: parveenzaman1989@gmail.com*

ABSTRACT

Moth bean is commonly grown in arid and semiarid regions of India. It is commonly called mat bean, moth bean, matki, Turkish gram, or dew bean. Moth bean also called matki, math, and dew bean has been known for high degree of adaptation in rainfed arid situation due to tolerance to drought and high temperature. The pods, sprouts, and protein-rich seeds of this crop are commonly consumed in India. Moth is a creeping annual which can be used as a green manure and as a fodder. Moth bean is drought tolerant and high protein content and is a potential crop choice for semi-arid Africa; a lack of management knowledge and the difficulty of harvest due to its density and creeping nature could make its spread to other parts of the world difficult.

10.1 BOTANICAL CLASSIFICATION

Kingdom: Plantae
Order: Fabales
Family: Fabaceae
Subfamily: Faboideae

Genus: *Vigna*
Species: *aconitifolia* (Jacq.)
Binomial name: *Vigna aconitifolia* (Jacq.)

10.2 BOTANICAL NAME

Vigna aconitifolia (Jacq.), Marechal (Synonymous: *Phaseolus aconitifolius*)

10.3 COMMON NAME

Moth or dew bean (Hindi—moth; Tamil—nari payaru; Telegu—mitikalu)

10.4 INTRODUCTION

Moth bean is a drought-resistant legume, commonly grown in arid and semiarid regions of India. It is commonly called mat bean, moth bean, matki, Turkish gram, or dew bean. Moth bean also called matki, math, and dew bean (*Vigna aconitfolia*) has been known for high degree of adaptation in rain-fed arid situation due to tolerance to drought and high temperature. The pods, sprouts, and protein-rich seeds of this crop are commonly consumed in India. Moth bean can be grown on many soil types and can also act as a pasture legume. Moth is a creeping annual which can be used as a green manure and as a fodder. Straw is a palatable fodder for livestock and dairy animals. It is used basically for fodder purposes and conserving soil and soil moisture. However, during one and half decade, need-based and deliberate attempts have been made to make this drought hardy crop more productive and adaptive in harsher and more hostile situations. In view of achieving the same, alternation in plant type and curtailment in maturity have been achieved up to desired level. While its drought tolerance and high protein content could make moth bean a potential crop choice for semiarid Africa, a lack of management knowledge and the difficulty of harvest due to its density and creeping nature could make its spread to other parts of the world difficult. The low-lying soil cover of the crop helps to prevent soil erosion and moisture loss (Sathe and Venkatachalam, 2007).

10.5 DESCRIPTION

Belonging to the family Fabaceae (subfamily Papilionaceae), the moth bean is a herbaceous creeping annual that creates a low-lying soil cover when it is fully grown (Sathe and Venkatachalam, 2007). Its stem can grow up to 40 cm in height, with its hairy and dense-packed branches. The leaves are alternate and trifoliate type. Leaflets are 5–10-cm long, deeply three lobed. The terminal leaf is divided into five acuminate lobes and the lateral leaflets are divided into three acuminate lobes. The roots of moth bean plants are well-developed taproot system and are highly responsive in water deficit conditions. In sufficient moisture conditions, plants spread its roots with nodulation. The nodulation forms in entire roots as like beads or as in cluster. The inflorescence is axillary and long peduncled. Yellow flowers develop into a brown pod 2.5–5 cm in length, which holds four to nine seeds inside. The rectangular seeds exist in a variety of colors, including yellow-brown, whitish green, and mottled with black. Flowers open generally in the morning. Moth bean is a self-pollinated crop. The moth bean pod is linear, buff to yellowish brown in color, about 3–36-cm long, and 5-mm wide with short stiff bristles. Seeds are yellow to brown or mottled black, somewhat reniform in shape with rounded. The seed germination is epigeal. Moth beans are short-day plants. The chromosome number is $2n = 10$.

10.6 ORIGIN AND DISTRIBUTION

Moth bean grows wild in India, Pakistan, and Burma. It is extended from the Himalayas in the north to Sri Lanka in the south. India is considered as the original home of moth bean. It is widely grown in India and the Far East, particularly Thailand. *Phaseolus aconitifolius* grows wild in India. It has reached China, Africa, and Southern United States where it is raised to some extent. Pakistan and Sri Lanka also grow this crop. The wild species and cultivated species of *P. aconitifolius* under Indian center of origin. Moth bean is cultivated as a crop only in India and that the wild original form of it occurs in both India and Ceylon. Moth bean can be found under cultivation throughout Southeast Asia and Middle East. They are also sometimes cultivated in the Mediterranean in places like Italy.

10.7 ACREAGE AND PRODUCTION

In Indian subcontinent, its cultivation is concentrated in northwestern desert region adjoining Indo-Pakistan border. Moth bean cultivation in other countries like Sri Lanka, Burma, South China, and Malaysia is only in limited scale. It was introduced into the United States, specially California and Texas. In India, Moth bean mainly grows in Rajasthan which constitutes about 93% of the total area and 85% of the total area in the country. The states like Maharashtra, Gujarat, and Jammu and Kashmir also grows it in limited scale.

10.8 CLIMATIC REQUIREMENT

Moth bean grows in tropical and subtropical areas of lower altitudes. It is a *kharif* season crop, cultivated in arid and semiarid regions. The crop prefers best hot and dry climates with annual rainfall which are about 60–80 cm.

10.9 SOIL AND ITS PREPARATION

Moth bean can grow in all types of soil. It is adapted in poor barren light textured soils. However, it prefers sandy loam soil for its cultivation. Moth bean is not suitable in saline and alkaline soil for its commercial cultivation. Moth bean can withstand in low-moisture retention capacity light sandy soil. In heavy soil, drainage facility should be there for draining excess water.

Moth bean can grow in minimum land preparation. Before plowing, well-decomposed farm yard manure or organic manure can broadcast for higher green fodder production. Generally, field is prepared by one plowing with mold board plow followed by one or two cross-harrowing. After final plowing, land is planked by wooden or iron plank for well distribution of the seeds. To ensure good germination presowing irrigation is given.

10.10 VARIETIES

Local varieties are grown which are of spreading type, indeterminate and viny, late in maturity prone to shuttering, and highly susceptible to yellow mosaic diseases. Little breeding work has been completed on the moth bean, but researchers have found that there is substantial genetic variation between moth bean germplasms (Yogeesh et al., 2012). The National Bureau of Plant Genetic resources in New Delhi, India, conserved more than 1000 accessions. Some improved cultivars such as CZM-2, CZM-3, "RMO-40" and "RMO-105" are available in India. Some moth varieties which are high forage yield potential as well as grain yield potential these are CAZRI Moth-2, CAZRI Moth-3, RMO-435, etc. Important characteristics of some varieties are given below:

1. *Jadia*: It is developed by local selection in Rajasthan maturing in 90 days yielded about 5–8 q/ha and green fodder yield about 400–600 q/ha.
2. *Baleshwar 12*: A selection from local germplasm; maturity period 110–115 days; seed weight 25 g/1000 seeds; average yield of 6–8 q/ha, and green fodder yield about 300–500 q/ha.
3. *Jawel*: Identified in 1983; maturity of 90 days; resistant to yellow mosaic virus, small seeds; light brown color with test weight of 30 g/1000 seeds, average yield of 7 q/ha, and green fodder yield about 300–500 q/ha.
4. *RMO-40*: By selection from Bikaner local variety. Most suitable for moth bean growing areas It is gaining popularity day by day. Green fodder yield about 400–600 q/ha.
5. *T 3*: A selection from local bulk in Meerut (Uttar Pradesh), India, as a good fodder variety.
6. *Type 1*: Developed by selection from Meerut local in 1967; plants spreading type; maturity period of 120 days; medium-sized seed; shining seed coat; brown color, and test weight of 190 g/1000 seeds, average grain yield 10–14 q/ha and green fodder yield about 400–600 q/ha.
7. *G Moth 1*: A selection from local germplasm of Gujarat; maturity period of about 90 days, and green fodder yield about 200–400 q/ha.
8. *IPCMO-0880*: A selection from Jhunjhunu local, released in 1989 takes 90–100 days to maturity. It is recommended or Rajasthan and Gujarat has a yield of 5 q/ha.
9. *IPCMO-912*: A selection from Sikar local, released in 1994 for Rajasthan. Maturing in 75–80 days has tolerance to YMV and bacterial blight.

10.11 SOWING TIME

Moth bean is usually sown with onset of monsoon in June or July. It is mainly sown in July when the monsoon rains are usually received in north and northwest India.

10.12 SEED TREATMENT AND INOCULATION

To control white grub, 25 kg phorate 10 G should be applied in the furrows where termite problem is common. Seeds should be inoculated with bacteria bradyrhizobium species before sowing for fixation of atmospheric nitrogen. Apply 20 g bradyrhizobium for 1 kg of seed before sowing for better nodulation.

10.13 SEED RATE AND SOWING METHOD

The seed rate is 15–20 kg for the sole crop and 2–10 kg/ha for intercrop. For fodder crops, seed rate generally gives higher than the gain production. For fodder production, the optimum seed rate 20–25 kg/ha is recommended.

10.14 CROPPING SYSTEMS

Moth bean can be included in various cropping systems in the *kharif* season. The crop is generally raised as dryland crop. Moth bean may be grown either frequently mixed or intercropped with pearl millet, guar, cowpea, pigeon pea, maize, jowar, bajra, or cotton.

- Some of the important rotation of the moth bean is given below:
- Sorghum–moth bean–barley
- Moth bean–pearl millet–mustard
- Moth bean–gram
- Maize + moth bean–mustard
- Moth bean–toria–wheat–green gram

10.15 MANURES AND FERTILIZERS

Application of organic manure, nitrogen, and phosphorus has been found to give higher green fodder yield. Farm yard manure or compost at 5–10 t/ha is desired for the crop under rainfed conditions. Apply 15–20 kg/ha N and 40–45 kg/ha P_2O_5 at the time sowing as a basal.

10.16 WATER MANAGEMENT

Moth bean is generally grown under rain-fed conditions under residual soil moisture conditions. It can stand the drought conditions quite well. *Kharif* season crop is given one premonsoon irrigation at the time of seedbed preparation to ensure good germination. It does not require any irrigation until an aberrant weather or drought occurs. In drought condition, the crop is irrigated as and when it needs. Two irrigations are required in addition to a presowing irrigation. One irrigation can be given at flowering stage and the second at grain development stage.

10.17 INTERCULTURAL OPERATION AND WEED MANAGEMENT

Broadcasted moth bean fodder crop is generally not practiced intercultural operation and weeding. Crop weed competition is more common in early stage of crop growth (25–35 days after sowing). Normally in most growing areas of Southeast Asia, weeds are controlled by hand weeding. This crop often is considered a low input crop with lower returns and therefore weeding may not take place.

10.18 INSECT-PEST AND DISEASE MANAGEMENT

Moth bean is infested by mungbean yellow mosaic virus, for which silverleaf whitfly is the vector. Root rot and seedling blight from *Macrophomina phaseolina* also cause damage, as well as some Striga species and the nematode *Meloidogyne incognita*. There are some resistant cultivars to these pests and diseases.

10.19 HARVESTING

The leaves turn yellow and the pods turn gray when mature. Pod splitting, which leads to premature shattering of seeds, is common in small types found on the Indian subcontinent. There, the plants are pulled out by hand or cut with a sickle near the base. The plants are then stacked and allowed to dry in the field or on the threshing floor for 7–8 days. The plants are spread out on the threshing floor and beaten with sticks. The seeds are more resistant to damage during harvesting operations than are field peas. It is a common practice to use cattle to help thresh the pods by trampling under the feet. The seed is then winnowed and cleaned. The seeds may be dried for several days before being stored. The straw and chaff are used for cattle feed, generally mixed with rice straw. The growers value the straw as a feed stuff as it is an important source of protein. In many cases, the value of the straw can be as great as that of the grain and thus is a major consideration for the farmer in the production of this pulse.

10.20 YIELD

Although the small-seeded types are desired for "utera" production, it can be seen that they are not necessarily higher yielding under these conditions. The amount of seed that is used per area of production does become important as the small-seeded types need significantly less than do large-seeded types to produce the same stand. As a rainfed crop, it shows green forage yield potentiality of 200 q/ha.

10.21 SEED PRODUCTION

Moth bean varieties have different maturity grows in 72–75 days, 65–67, 60–62, and 57–58 days, respectively, with semi spreading, semi-erect and erect-upright plant type. These varieties suit to 450–500, 300–450, 150–300, and 130–150 mm rainfall during crop season, respectively, with reasonable grain yield potential from 500 to 1400 kg/ha.

10.22 NUTRITIVE VALUE

The moth seed contains starch 48.0–52.3%, protein 25.6–28.4%, acid detergent fiber 4.3–7.3%, ash 2.9–4.6%, and fat 0.58–0.8%. Around 100 g of raw, uncooked moth bean seeds contain 343 calories, 23 g of protein, 62 g of carbohydrate, and 1.6 g of fat. As is the case with other legumes, this pulse does contain antinutritional factors that limit available protein. However, research has shown that the moth bean contains considerably less of these factors compared with other legume grains, making it a more beneficial choice for consumption. Soaking and cooking moth beans before consumption helps to break down antinutritional factors and makes the protein more digestible (Khokhar and Chauhan, 1986).

10.23 UTILIZATION

Whole or split moth bean seeds can be cooked or fried. In India, particularly in the state of Maharashtra, moth beans are sprouted before cooking and used for making usal. They can be used for breakfast or other meals. The usal is essential part of the popular dish missal pav (Nimkar et al., 2005). Fried splits make up a ready-to-eat traditional namkeen, or savory dry snack, in India called *dalmoth* and can be used to make traditional dal. The moth bean pods can be boiled and eaten, and seeds can be ground into flour that is used for another traditional namkeen called bhujia. It is believed that consumption of the seeds can help treat a fever. Moth bean is also consumed as a forage crop by animals.

KEYWORDS

- **moth**
- **agronomic management**
- **fodder yield**
- **nutritive value**
- **utilization**

REFERENCES

Khokhar, S.; Chauhan, B. M. Antinutritional Factors in Moth Bean (*Vigna aconitifolia*): Varietal Differences and Effects of Methods of Domestic Processing and Cooking. *J. Food Sci.* **1986,** *51* (3), 591–594.

Nimkar, P. M.; Mandwe, D. S.; Dudhe, R. M. Properties of Moth Gram. *Biosyst. Eng.* **2005,** *91* (2), 183–189.

Sathe, S. K.; Venkatachalam, M. Fractionation and Biochemical Characterization of Moth Bean (*Vigna aconitifolia*) Proteins. *LWT-Food Sci. Technol.* **2007,** *40* (4), 600–610.

Yogeesh, L. N.; Viswanatha, K. P.; Ravi, B. A.; Gangaprasad, S. Genetic Variability Studies in Moth Bean Germplasm for Seed Yield and Its Attributing Characters. *Electron. J. Pl. Breed.* **2012,** *3* (1), 671–675.

CHAPTER 11

SOYBEAN (BHATMAN)

SAGAR MAITRA*

Department of Agronomy, M. S. Swaminathan School of Agriculture, Centurion University of Technology and Management, Paralakhemundi 761211, Odisha, India

**E-mail: sagarmaitra69@gmail.com*

ABSTRACT

Soybean is called a "wonder crop" and "golden bean" due to its high protein and fat. Soybean is native to Asia that is currently grown worldwide. The foliage is very palatable to cattle and has a high nutritive value and good digestibility. Soybean forage is much valued in wildlife management as it is also palatable to deer. Similar to most legumes, soybean forage tends to be high in protein and low in fiber making. Soybean green fodder yielded about 200–110 q/ha under good management practices. Soybean forage is rich in crude protein (18.2–20.3%) and with low in fiber.

11.1 BOTANICAL CLASSIFICATION

Kingdom: Plantae
Order: Fabales
Family: Fabaceae
Subfamily: Faboideae (Papilionoidae)
Genus: *Glycine*
Species: *max* (L.)
Binomial name: *Glycine max* (Linn.)

11.6.1 MANCHURIAN CLASSIFICATION

This classification is based on the color of seed coat and embryo.

1. *Yellow group*: This group is also into three subgroups.
   a) Yellow seeds with light hilum
   b) Yellow seeds with golden hilum
   c) Yellow seeds with brown hilum
2. *Black group:* It has three divisions.
   a) Large black seeds
   b) Flat black seeds
   c) Small black seeds
3. *Green group:* It is further divided into two groups.
   a) Epidermis is green but embryo is yellow
   b) Epidermis as well as embryo is green

11.6.2 MARTIN'S CLASSIFICATION

This classification is based on the shape of the seed and divided into following three categories.

1. *Soya eliptica*, seed is of egg type.
2. *Soy spherical*, seed is round shaped.
3. *Soy compress*, seed is pressed type.

11.6.3 HERTZ CLASSIFICATION

This classification is based on the shape of pods and classified soybean into following two groups:

1. Soya platycarpa
2. Soya tumida

11.6.4 US CLASSIFICATION

The classification is based on the maturity period and hilum color of varieties and these are grouped into 10.

11.7 ORIGIN AND DISTRIBUTION

Generally, it is believed that soybean is a native of eastern Asia. In his work, Nagata (1960) described that the distribution of *Glycine ussuriensis*, the progenitor of *G. max*, was found in north central region of China. *G. ussuriensis* was grown in Korea, Taiwan, Japan, and Northern Province of China and adjacent area of Russia. However, Chang (1989) claimed that yellow river valley of China is the origin of soybean. It was introduced in the United States in the 18th century and systematic breeding started there during middle of 20th century. After that, this forage crop has been transformed to a highly productive and nutritive crop. The United States is the largest producer of soybean followed by Brazil and Argentina.

11.8 PLANT CHARACTERISTICS

The cultivated soybean is an erect and bushy annual plant. It has diversity in different morphological characters. The plant height may vary from 0.3 to 2.0 m and having tap root system with many branches. The major part of active root remains up to the soil depth of 60 cm. Nodules are formed in the root as *Bradyrhizobium japonicum* bacterium remains in root symbiotically and fixes nitrogen biologically. The root system becomes extensively nodulated during maturity.

The mature stem is largely pithy tissue and woody in basal part. Stems are formed as a result of hypocotyll elongation of the seed. The height of the stem and branching habit depends on varieties. The forage types are profusely branched with long prostate or horizontal creeping branches, whereas they are short and erect in grain types. In most of the varieties, stems are found hairy.

Soybean has four types of leaves—(1) cotyledons or seed leaves, (2) two primary leaves, (3) trifoliate leaves, and (4) prophyllus. Seed leaves emerge with seedlings. The primary leaves and seed leaves are arranged opposite while trifoliate leaves are alternatively arranged on the stem. The leaflet of trifoliate leaves has entire margins and varies from oblong to ovate or lanceolate in shape. The fourth type, prophyllus leaves are very tiny paired and those are present at the base of each lateral branch.

Two types of inflorescence have been found in case of soybean. The first type appears in indeterminate stems where terminal buds grow and produce axillary raceme type of inflorescence. In this case, pods are evenly distributed on the branches and number of pods reduces toward the top of the stem. The second type of inflorescence appears in the determinate stems which contain both axillary and terminal racemes and pods are found in dense clusters along the stems. Floral primordial is initiated within 3 weeks and flower begins 6–8 weeks after germination. Flowering may continue for 3–4 weeks. Flower of soybean is self-fertilized and pollination may occur the day before full opening of the flower even it takes place within the bud. Soybean pods are straight or slightly curved. The length of pod may vary from 2 to 7 cm depending on varieties. In most of the varieties, pods are hairy. Seeds develop inside the pod and fully developed seeds are formed after 35–45 days of blooming. As per the varietal character, the size, shape, and color of seed may vary.

11.9 CLIMATIC REQUIREMENT

Soybean grows well in warm and moist climate. The maximum and minimum temperature requirements for germination of soybean are 5 and 40°C, respectively; however, optimum temperature for rapid germination is 40°C. Temperature of 18°C or less does not permit pod set. Seed size reaches maximum when the crop is grown at 27°C and number of pods goes maximum at the temperature of 30°C. Temperature goes below 24°C, if the flowering delays by 2–3 days and flowering greatly influenced at temperature 10°C or less. Pod abscission may be caused at 40–46°C.

Soybean requires 450–900 mm rainfall; however, 550–600 mm rain is considered as good enough for successful rising of crop (Billore et al., 2004). Prior to flowering, the crop can tolerate drought because of its deep root system, but after flowering to pod development stage, water stress is not desirable which may reduce yield.

11.10 SOIL AND ITS PREPARATION

As soybean has wider adaptability, it can be grown in different types of soil, namely, vertisols, ultisols, and entisols. To get better yield, it is necessary to raise soybean in well-drained, sandy-loam to clay loam, reasonable depth (up to 1 m or more), rich in organic carbon with near neutral pH (6.5–7.5). Highly acidic or saline and poor drained soil should be avoided.

Good tilth condition is required for sowing of soybean and for that purpose one deep plowing and followed by 2–3 cross harrowing or cultivation is required. In pulverized soil, crop performs better. The field should be leveled by planking and a gentle gradient should be provided to ensure good drainage for monsoon crop of soybean in tropics and subtropics. At the time of land preparation, stubbles of previous crop and weeds should be removed. If there is any shortage of soil moisture, on presowing, irrigation may be provided for optimum stand.

11.11 VARIETIES

In India, a number of varieties are cultivated in different soybean-growing states. Northern hills zone varieties are VLS 63, VLS 59, VLS 47, VLS 21, Palam Soya, Harit Soya, and Bragg. Northern plains varieties are PS 1347, PS 1241, PS 1211, PS 1092, PS 1024, PS 1029, PK 472, PK 416, SLS 688, Pusa 9814, Pusa 9712, and Bragg. Central zone varieties are JS 93-05, JS 335, JS 95-60, JS 90-41, Ahilya 4 (NRC 37), Ahilya 3 (NRC7), Indra Soya 9, MAUS 81, MAUS 61-2, MAUS 47, and Bragg. Southern zone soybean varieties are Pratap Soya 2, TAMS 98-21, TAMS 38, Phule Kalyani, MAUS 61, MAUS 32, MAUS 2, MACS 450, MACS 124, LSb-1, PS 1029, Bragg, etc. Northeastern zone varieties are Pratap Soya 2, Pratap Soya 1, JS 97-52, JS 80-21, MAUS 71, Indra Soya 9, and Bragg (Agarwal et al., 2010).

11.12 CROPPING SYSTEMS: CROP SEQUENCE, INTERCROPPING, AND CROP MIXTURE

Soybean is having wider adaptability; it can be cultivated under different cropping systems like sequential cropping, companion cropping, intercropping, and mixed cropping. It has also some qualities which made the crop suitable for different cropping systems like tolerance to shed and drought, compact canopy, can fix nitrogen biologically and benefit the succeeding crop or nonlegume crop in sequence or in association when raised as intercrop or mixed crop and also a remunerative crop. In India, different types of crop sequence are adopted at different zones and intercropping of soybean is very much common with sorghum, maize, pearl millet, finger millet, sugarcane, cotton, pigeon pea, etc (Bhatnagar and Joshi, 1999). Some of the important soybean-based crop sequences are given below:

1. Soybean–wheat
2. Soybean–chick pea
3. Soybean–wheat–corn–fodder
4. Soybean–potato
5. Soybean–garlic/potato–wheat
6. Soybean–rapeseed/mustard
7. Soybean–pigeonpea
8. Soybean–safflower
9. Soybean–sorghum
10. Soybean–linseed

Some of the important soybean-based intercropping/mixed cropping are given below:

1. Soybean + pigeonpea
2. Soybean + sorghum
3. Soybean + groundnut
4. Soybean + pearl millet
5. Soybean + cotton
6. Soybean + sugarcane

The most suitable crop sequence may be soybean followed by nonlegume crops. In case of intercropping/mixed cropping soybean and nonlegume, combination is beneficial in terms of productivity of crops as well as restoration of soil fertility. In an experiment conducted in ultisols of West Bengal, India, it has been observed that intercropping of finger millet and soybean (4:1) showed higher monetary advantages than intercropping of finger millet and green gram or groundnut with the same row proportion (Maitra et al., 2000).

11.13 SOWING TIME

In India, soybean is mostly a rainy season crop and sown between the middle of June to July. However, there is regional variation in sowing dates. In Northern hills, sowing time is last week of May to end of June and in northern plains is middle of June to first week of July. In east and northeast parts, sowing time is middle of June to mid of July, whereas central and southern parts sowing time of soybean is middle of June to end of July.

Under irrigated conditions, spring crop of soybean may be raised in northeastern states where sowing is done between mid of February and mid of March. Too early sowing of soybean in rainy season ensures greater vegetative growth and late sowing results poor canopy and both of these are nonbeneficial in terms of seed production.

11.14 SEED TREATMENT AND INOCULATION

To protect the crop from pests and diseases, seeds are treated by fungicides and insecticides. In soybean, seeds are also inoculated by *Rhizobium japonicum* to ensure nodulation and biological nitrogen fixation. To treat and inoculate the seed, a sequence is to be maintained. It is advised to treat seeds by insecticide first and that may be with Imidachloprid 600 FS (*Gaucho*) @ 75 g a.i. (or 111-mL formulation) per 100 kg of soybean seed to avoid the attack of *jassids* and other sucking pests. After providing a gap of 1 week, the insecticide-treated seeds are further treated by fungicides like Thiram or captan @ 3 g/kg of seed to avoid colar rot and kept it for 1 week. After that, the insecticide and fungicide-treated seeds are inoculated by the biofertilizer *R. japonicum* and sown to the field.

11.15 SEED RATE AND SOWING METHOD

Seed rate of soybean greatly depends on size and weight of seed, germination percentage, and date of sowing. In case of small seeded variety (seed index < 10 g), 60–65 kg seed will be required for sowing of 1 ha of land. However, seed rate for medium seeded variety (seed index < 11–12 g) and large seeded variety (seed index > 12 g) are 70–75 and 80–85 kg/ha, respectively. Seeds are sown in rows with a depth of 3–5 cm. Row × plant spacing is generally maintained 45 cm × 5 cm in rainy season crop and 30 cm × 5 cm for spring crop. In rainy season crop, for late sowing, closer spacing is advised which requires more seeds.

11.16 NUTRIENT MANAGEMENT

Soybean is a high-nutrient removing crop. The dose of $N:P_2O_5:K_2O$ is 20, 60, and 20 kg/ha, respectively, in all zone/parts of India and sulfur 20 kg/ha at the time of land preparation.

Therefore, proper nutrient management should be adopted for sustainability of soybean production. The amount of nutrients to be applied actually depends on the soil fertility status and cropping system to be adopted. Soybean responds to different micronutrients like zinc, boron, and molybdenum which improve oil as well as protein content and productivity. In general, 5–10 kg zinc/ha is recommended as basal dose or foliar application of zinc sulfate (0.5%) is done. To provide boron, basal dose varies from 0.5 to 2.0 kg/ha or foliar spray of sodium tetraborate (0.5%) is recommended twice at vegetative and reproductive stages. Similarly, ammonium molybdate can be added in deficient soil as basal dose of 0.8 kg/ha. However, micronutrients should be added as basal dose in every 2–3 years. Application of farmyard manure @ 10 t/ha is sufficient to meet the requirement of micronutrients rather than application of individual elements (Tandon, 2002).

11.17 WATER MANAGEMENT

Soybean in India is cultivated as rainy season crop and it is mostly rainfed; however, supplemental irrigation may be required if any dry prevails mainly during reproductive stage. Generally, seedling, flowering, and pod filling are considered as the critical stages of irrigation when scarcity of soil moisture is not desirable to get satisfactory yield of soybean. The crop requires more water during flowering to pod formation stage, and during this stage, daily water requirement is 0.75 cm. The vegetative stage of soybean utilizes only 11–30% of total water requirement and majority of water requires during reproductive stage. Soybean requires 50 cm water for proper growth and development of the crop, and as a rainy season crop in India, it gets almost the same from natural precipitation. But for spring crop assured, irrigation is essential and the crop needs 4–6 irrigations, 4–5-cm depth each.

Moreover, soybean cannot withstand to waterlogging or excess soil moisture conditions. Therefore, it is essential to make necessary arrangements for proper drainage during rainy season.

11.18 WEED MANAGEMENT

Soybean requires two hand weeding at 20–11 and 40–45 days after sowing. If there is severe weed infestation, particularly in rain-fed uplands in India during rainy season, it is advisable to go for chemical control measures with herbicides. Calmazone 50 EC @ 1.5 kg a.i./ha as preemergence or Imazethapyr 10 SL @ 0.1 kg a.i./ha as postemergence may be applied to control both monocot and dicot weeds of soybean. Besides, other options of integrated weed management may also be considered like sowing of weed-free certified seed, deep summer tillage, destruction of weeds at the time of soil preparation, adoption of proper crop rotation, maintaining optimum crop stand, and judicious fertilizer application as well as integrated nutrient management. However, adoption of these management practices along with one preemergence or postemergence application of herbicide and one intercultural operation at 11–30 days after sowing is found effective and economically viable for production of soybean in India.

11.19 DISEASE AND INSECT-PEST MANAGEMENT

Plant protection is an important aspect to get target yield. The crop may suffer from different diseases and insect-pests which may be managed by adopting integrated approach.

11.19.1 DISEASES

Different fungal, bacterial, and viral diseases are found in soybeans and if they are not managed properly, crop yield is hampered.

11.19.1.1 CHARCOAL ROT

It is caused by *Rhizoctonia bataticola* (syn. *Macrophomina phaseolina*). Reddish brown to black necrotic lesions develop at the crown. This hampers translocation of fluid and plants may die. It can be managed by cultural operations like summer tillage, crop rotation, nutrient management, and presowing seed treatment by thiram @ 1.5 g + carbendazim

@ 1.5 g/kg of seed. If crop is infected, spraying of copper oxichloride @ 0.4% is beneficial.

11.19.1.2 AERIAL BLIGHT

It is caused by *Rhizoctonia solani*. The leaves, stems, and pods are affected. The first symptoms appear on lower leaves, the water soaked lesions turn in brown to black. Complete leaf may be blighted slowly. Infected leaf may drop and adhere to stem and pods. Seed treatment with fungicides, summer tillage, and crop rotation is beneficial to manage the disease. In infected crop, mancozeb (0.11%) can be sprayed. Some tolerant varieties in India like PK 416, PS 564, and PS 1042 may be grown (Gupta, 2001).

11.19.1.3 ALTERNARIA LEAF SPOT

It is caused by *Alternaria alternata*. The symptom of this disease is brown necrotic spots and concentric rings in all above ground parts of the plant. Dry and moist weather is favorable for spread of the disease. To avoid the disease, it is essential to sow disease-free seed with prior treatment with fungicides. Mancozeb (0.11%) can be sprayed as protective measure.

11.19.1.4 ANTHRACNOSE

It is caused by *Colletrotichum trancatum*. Symptoms of the disease may appear on leaves, pods, stems, and petioles with irregular brown to black spots. Infected seeds may turn into moldy, dark brown, and shriveled. Cultivation of disease-free seed and seed treatment is advised to manage the disease. However, in infected field, foliar spray of mancozeb (0.11%) is also effective.

11.19.1.5 RUST

It is caused by *Phakopsora pachyrhizi* and *Phakopsora meibomiae*. It is one of the most destructive diseases of the soybean and found mainly in tropical and subtropical countries. Initially, small gray brown spots appear on the leaves of infected plants which enlarge later. Ultimately, leaves turn rust like brown color. It may cause 30–100% yield loss. Foliar spray of propiconazole or hexaconazole for two to three times is effective to control the disease.

11.19.1.6 BACTERIAL PUSTULES

It is caused by *Xanthomonas phaseoli* var. *sojensis*. At an early growth stage, minute leaf spots are grown on either side of leaf. Later red brown colored pustule develops in the center of the spots. The best management practices of the disease are use of disease-free seed, crop rotation, and destruction of residues of infected field. However, spraying of copper oxychloride (0.4%) + streptocycline (0.01%) is found beneficial to protect the crop.

11.19.1.7 BACTERIAL BLIGHT

It is caused by *Pseudomonas glycinea*. The symptom of the disease is appearance of small, angular translucent, water-soaked yellow spots on leaves, stems, petiole, and pods. These spots turn brown to black and may result in fall of leaves. Summer plowing and disease-free seed sowing help in preventing the disease. Spraying of copper oxychloride (0.4%) + streptocycline (0.01%) is also beneficial as protective measure.

11.19.1.8 YELLOW MOSAIC

Vector of this viral mosaic is whitefly (*Bemisia tabaci*). The symptom of the disease mainly appears on leaves, mixture of yellow and green mosaic like appearance. Growth of affected plant is stunted. The vectors may be controlled by spraying of chemical insecticides like thiamethoxam 11 WG (100 g/ha) or methyl demeton 11 EC (1.11 L/ha) or quinalphos 11 EC (1.2 L/ha). It is not a seed-borne disease. There are some resistant varieties in India like PK 416, PS 564, PS 1042, etc. which may be grown in yellow mosaic prone areas.

11.19.1.9 SOYBEAN MOSAIC

This is a seed-borne disease, and in crop field, it is transmitted by aphids. The affected plants become stunted with darker and reduced leaves. Yield of soybean is affected greatly. Use of disease-free seed is essential to prevent the disease. To control vector, foliar spray of thiamethoxam 11 WG (100 g/ha) or methyl demeton 11 EC (1.11 L/ha) or quinalphos 11 EC (1.2 L/ha) can be done.

11.19.2 INSECT-PESTS

11.19.2.1 HAIRY CATERPILLAR (Spilosoma obliqua)

This polyphagus pest in cluster attacks the crop and feeds on chlorophyll of leaves and thus leaves appear with only veins. It can be controlled by spraying quinalphos 11 EC (@ 1.11 L/ha) or chloropyriphos 20 EC (@ 1.11 L/ha).

11.19.2.2 STEM FLY (Melanagromyza sojae)

The adults of stem fly feed on the leaves by making small punctures, which ultimately turn into white spots. The maggots feed on the leaves and move toward the center of the stem. The infested plants dry and leaves droop. Chemically, this insect may be controlled by spraying thiamethoxam 11 WG (100 g/ha) or triazophos 40 EC (@ 0.8 L/ha).

11.19.2.3 WHITEFLY (B. tabaci)

It is one of the serious pests of soybean which not only sucks the sap from leaves but also transmits yellow mosaic virus. To control it, foliar spray of thiamethoxam 11 WG (100 g/ha) or methyl demeton 11 EC (1.11 L/ha) or quinalphos 11 EC (1.2 L/ha) can be done.

11.19.2.4 LEAF ROLLER (Hedylepta indicata)

The larvae of this Lepidopteran insect fold leaves around themselves and feed on mesophyll resulting in an intact papery skeleton of folded leaves.

It can be controlled by spraying quinalphos 11 EC (@ 1.11 L/ha) or chloropyriphos 20 EC (@ 1.11 L/ha).

11.19.2.5 GRIDDLE BEETLE (Obereopsis brevis)

It is one of the harmful pests of soybean. The female insect, after mating, makes two parallel girdles usually on the petiole or main stem or side branches and bores several holes in the girdle and lays eggs. If petioles are girdled, the leaves are curled and leaflets become dry.

11.19.2.6 TOBACCO CATERPILLAR (Spodoptera litura)

Newly hatched larvae in cluster feed gregariously on chlorophyll of leaves. In later instars, these larvae attack neighboring plants and damage leaves by cutting big holes. The pest may be controlled chemically by spraying chloropyriphos 20 EC (@ 1.11 L/ha) or methomyl 40 SP (@ 1 kg/ha) or Lambda cyhalothrin 5 EC (@ 300 mL/ha).

11.19.2.7 GREEN SEMILOOPER (Diachrysia orichalcea)

Young larvae of green semilooper are pale greenish-white and they feed singly on soft tissues, leaving veins of leaves. To control this harmful insect, quinalphos 11 EC (@ 1.11 L/ha) or thiamethoxam 11 WG (100 g/ha) may be sprayed.

11.19.2.8 GRAM POD BORER (Helicoverpa armigera)

It is a polyphagus pest and one of the major pests of soybean. The caterpillar feeds on grains and reduces the yield drastically. In later instars, it is difficult to control. Foliar spray of methomyl 40 SP (@ 1 kg/ha) or Lambda cyhalothrin 5 EC (@ 300 mL/ha) is beneficial to control the pest. To manage the insect-pests of soybean, one should adopt all suitable tools of integrated pest management to maintain pest population dynamics below the threshold level. These measures are choice of pest tolerant varieties, deep summer plowing to destroy the pupae remained in the soil, installation of light trap for (sensitive pests) and pheromone trap (for *S. litura* and *H.*

armigera), yellow sticky trap (for *B. tabaci*), sue of bird perches, spraying of biopesticides like *Bacillus thuringiensis* (*Bt*), and nuclear polyhedrosis virus for *S. litura* and *H. armigera*, mechanical removal of infested plant parts and collection and destruction of larvae of late instars.

11.20 HARVESTING

Timely harvesting is essential in soybean to get quality and quantity. Overmature crop may shatter seeds. The time of maturity depends on varietal characteristics and it ranges from 80 to 130 days. Soybean is harvested when the leaves start drooping and falling and color of pods turn into brown to black. At the time of harvesting, moisture content of grain may be 15–17%. Small farmers generally harvest the crop manually; however, combine harvester may be used for large areas. Seeds can be dried after threshing at 12% moisture for ideal storage.

11.21 YIELD

The yield of soybean depends greatly on duration of the varieties. Generally, short duration varieties in India produce seed yield of 2.0–2.5 t/ha and long-duration varieties yield 2.5–3.0 t/ha. However, 10–12 t/ha forage yield can also be harvested from soybean.

11.22 SEED PRODUCTION

The good quality soybean should be with 98% genetic purity and 70% germination. In soybean seed, inert material is restricted to 2% only. Besides, 1 kg good quality seed of soybean should not contain more than 10 numbers of other crop seed and 10 numbers of weed seeds. General agronomic managements should be adopted for seed production of soybean that has already been stated earlier. However, some steps are to be followed to get proper quality seeds.

The parent material, that is, nucleus/breeders/foundation seed, should be collected from the authentic source as approved by the seed certifying agency. It is normally self-pollinated crop and cross pollination by insects is less than 1%. But an isolation distance of 3 m from other soybean field is sufficient to maintain genetic purity. Proper plant protection measures should be adopted to avoid yellow mosaic disease. From 2 to 3 weeks after sowing, rouging of mosaic affected plants should be done and it will be continued until harvest. At flowering stage, offtype plants should be removed on the basis of plant characteristics and flower color. Final roughing should also be done at maturity on the basis of pod characteristics.

The seed crop of soybean should be harvested only on maturity and premature harvesting must be avoidable. After that, standard postharvest technologies should be followed. Care must be taken in the threshing floor to avoid mixing among other varieties of soybean or other crop seeds. The seeds obtained should be sun-dried and moisture content should be 12%. The seeds may be stored in gunny bags and kept on wooden racks in a well-ventilated and dry store.

11.23 UTILIZATION

As soybean is highly nutritious, different canned food and dry foods are made from soybean like soy beverage, soy milk and curd, ice cream, candy, soy nuts, cheese, nuggets, chunks, soy flour, and snacks as value-added products. The oil can be used as edible oil. Besides, there is several industrial use of soy oil as glycerin, explosives, varnish, printing ink, paint, soap, celluloids, and lecithin tocopherols (vitamin E) are produced. The remaining plant parts can be used as forage or making of silage and hay. In Europe, there is a restriction in using animal protein in animal feeds and this increased the use of soybean meals in animal diet. Globally, soybean is the major source of protein in animal feed and its use is growing steadily. In India also, it has enormous potential as an ingredient in poultry feed. About 85% of the world's soybeans are processed annually into soybean meal and oil. Approximately 98% of the soybean meal is crushed and further processed into animal feed with the balance used to make soy flour and proteins. Of the oil fraction, 95% is consumed as edible oil; the rest is used for industrial products such as fatty acids, soaps, and biodiesel.

11.24 NUTRITIVE VALUE

Soybean seeds supply 446 calories, 9% water, 11–30% carbohydrates, 20% total fat, and about 40% good quality protein and almost no starch (beneficial to diabetic patient) in each 100 g of raw product. The soybean oil is having 85% unsaturated fatty acids including 55% polyunsaturated fatty acid. It also contains 4–5% minerals and vitamins as well as antioxidants like ascorbic acid (9–10 mg/100 g sprouted soybean), beta-carotene (0.2 mg/100 g sprouted soybean), and about 0.3% isoflavones. Soybean forage is rich in crude protein (18.2–20.3%) and with low in fiber (Reddy, 2001). Soybean is called a "wonder crop" and "golden bean" for this reason.

FIGURE 11.1 (See color insert.) Soybean.

KEYWORDS

- **soybean**
- **origin**
- **morphology**
- **characteristics**
- **cultivation**
- **fodder yield**
- **nutritive value**

REFERENCES

Agarwal, D. K.; Hussain, S. M.; Ramteke, R.; Bhatia, V. S.; Srivastava, S. K. Soybean Varieties in India. Directorate of Soybean Research, ICAR: Indore, India, 2010; p 76.

Bhatnagar, P. S.; Joshi, O. P. Soybean in Cropping Systems in India. In *FAO Series on Integrated Crop Management*; FAO: Rome, Italy, 1999; Vol. 3, p 39.

Billore, S. D.; Vyas, A. K.; Joshi, O. P. Soybean Agronomy. In *Soybean Production and Improvement in India*; Singh, N. B., Chauhan, G. S., Vyas, A. K., Joshi, O. P., Eds.; NRC for Soybean, ICAR: Indore, India, 2004; pp 106–111.

Chang, R. Studies on Origin of Cultivated Soybean (*Glycine max* L./Merrill). *Oil Crops China* **1989,** *1*, 1–6.

Gupta, G. K. Management Strategies for Soybean Diseases in India. Souvenir. Harnessing the Soy Potential for Health and Wealth. *India Soy forum*, 2001; pp 38–45.

Maitra, S.; Ghosh, D. C.; Sounda, G.; Jana, P. K.; Roy, D. K. Productivity, Competition and Economics of Intercropping Legumes in Finger Millet (*Eleusine coracana*) at Different Fertility Levels. *Indian J. Agric. Sci.* **2000,** *70* (12), 824–828.

Nagata, T. Studies on Differentiation of Soybean in Japan and the World. *Men Ityogo Univ. Agric. Ser.* **1960,** *4*, 63–102.

Pal, S. S.; Gangwar, B. Nutrient Management in Oilseed Based Cropping System. *Fertil. News* **2004,** *49* (2), 37–45.

Reddy, V. R. In *Quality of Indian Soybean Meal for Livestock and Poultry*, Proceedings of India Soy Forum; Bhatnagar, P. S., Ed.; 2001; pp 208–211.

Tandon, H. L. S. *Nutrient Management Recommendations for Pulses and Oilseeds*; Fertilizer Development Consultation Organization: New Delhi, 2002.

CHAPTER 12

FENUGREEK (GREEK CLOVER)

PARVEEN ZAMAN[1], MD. HEDAYETULLAH[2*], and
KAJAL SENGUPTA[2]

[1]*Assistant Director of Agriculture, Pulse and Oilseed Research Sub-station, Department of Agriculture, Government of West Bengal, Beldanga 742133, Murshidabad, West Bengal, India*

[2]*Department of Agronomy, Bidhan Chandra Krishi Viswavidyalaya, Mohanpur, Kalyani 741235, Nadia, West Bengal, India*

[*]*Corresponding author. E-mail: heaye.bckv@gmail.com*

ABSTRACT

Fenugreek is an annual leguminous crop native to Asia and Southeast Europe. Fenugreek is a short-duration winter forage crop and can tolerate drought. The fact that fenugreek produces high-quality forages in all growth stages, does not create bloat problem in cattle, and contains animal growth-promoting substances makes it an attractive forage crop. Irrigated fenugreek crop gives 250–350 q/ha green fodder but rainfed crop gives 150–250 q/ha. The average seed yield of fenugreek is 6–8 q/ha. The chemical compositions of fenugreek are crude protein (3.6%), ether extract (3.5%), crude fiber (7.1%), nitrogen-free extract (9.5%), and ash (2%) at the flowering stage on dry weight basis. Coumarin-like compounds present in fenugreek may interfere with the activity and dosing of anticoagulants and antiplatelet drugs and cause birth defects in animals and it can pass through the placenta.

12.1 BOTANICAL CLASSIFICATION

Kingdom: Plantae
Order: Fabales

Family: Fabaceae
Subfamily: Faboideae
Genus: *Trigonella*
Species: *foenum graecum*
Binomial name: *Trigonella foenum-graecum* L.

12.2 BOTANICAL NAME

Trigonella foenum-graecum L.

12.3 COMMON NAME

Methi, metha, fenugreek, greek hay, greek clover, bockshornklee, alholva; it is known as *methi* in Hindi, Oriya, Bengali, Punjabi, and Urdu language. Fenugreek also known in other languages like *methya* in Marathi, *menthya* in Kannada, *vendayam* or *venthayam* in Tamil, *menthulu* in Telegu, *uluva* in Malayalam, and *Medhika* or *chandrika* in Sanskrit language.

12.4 INTRODUCTION

Fenugreek is a self-pollinating, annual leguminous crop native to Asia (probably India) and Southeast Europe and is cultivated worldwide. Fenugreek is a short-duration winter forage crop and can tolerate drought. The fact that fenugreek produces high-quality forages in all growth stages, does not create bloat problem in cattle, and contains animal growth-promoting substances makes it attractive forage crop (Acharya et al., 2008).

12.5 ORIGIN AND DISTRIBUTION

It is believed that fenugreek was known in India since 3000 years ago. Fenugreek is native to Asia (probably India) and Southeast Europe and is cultivated worldwide (Acharya et al., 2008). Its growth in the wild is reported from Kashmir, Punjab, and the Upper Gangetic plains. Fenugreek is also reported in ancient Egypt and India and later in Greece and Rome. A notable practice reported is its use as fumigant in incense burning in religious ceremonies in Egypt to spread "holy smoke." Other countries where it is grown are Argentina, Egypt, and Mediterranean region. Use of its seed as a spice and its leaves and tender pods as vegetable were also reported. Fenugreek is also used as cattle feed.

12.6 PRODUCTION

Major fenugreek-producing countries are Afghanistan, Pakistan, India, Iran, Nepal, Bangladesh, Argentina, Egypt, France, Spain, U.S.S.R., Ethiopia, Iraq, Turkey, and Morocco. The largest producer is India and fenugreek production in India is concentrated in the states of Rajasthan, Gujarat, Uttarakhand, Uttar Pradesh, Madhya Pradesh, Maharashtra, Haryana, Punjab, and West Bengal. Rajasthan is the leading producer of fenugreek, which produces about 80% of India's output (Parthasarathy et al., 2000).

12.7 PLANT CHARACTERISTICS

Fenugreek leaves are alternate and consist of three ovate leaflets. White flowers appear in early summer and develop into long slender green pods; mature brown pods contain 20 small yellow seeds. The stalks grow up to 2-ft (60-cm) tall.

12.8 CLIMATIC REQUIREMENT

Fenugreek favors cool and dry climate for its growth and development. Production happens well in areas where temperatures range from 8°C to 12°C and have an annual precipitation of 100–150 mm and a soil pH of 5.3–8.2. Fenugreek tolerates slightly saline soil but not acid soil. Growth is slow and weak in cold temperatures and wet soils. High humidity, freezing temperatures, and water stagnation are detrimental to this crop.

12.9 SOIL AND ITS PREPARATION

This crop prefers loamy and alluvial soil for its growth. It prefers neutral to slightly alkaline soil and is tolerant to saline conditions. The plant cannot survive under waterlogged situation. Fenugreek can withstand in low moisture retention capacity, light sandy soil. In heavy soil, drainage facility should be there for draining excess water. Fenugreek can grow in lightly land preparation, but for higher green fodder, production field should be well prepared. Before plowing, well-decomposed farmyard manure or organic manure can broadcast for higher green fodder production. Generally, field is prepared by one plowing with mold board plow followed by one or two cross-harrowing. After final plowing, land is planked by wooden or iron plank for the proper distribution of seeds. To ensure good germination, presowing irrigation is given (Chatterjee and Das, 1989).

12.10 VARIETIES

The variety FOS-8 has very good forage production ability. Some other fenugreek varieties have special characters like Lam selection 1 (tolerant to powdery mildew, root rot, caterpillars, and aphid), Pusa Early Bunching (resistant to downy mildew, rots), Kasuri methi (resistant to leaf minor), Rajendra Kranti (tolerant to leaf spot, suitable for intercropping), RMt-143 (moderately resistant to powdery mildew), etc. T-8 variety was developed by Anand Agriculture University (AAU), Anand in 1989 and is recommended for cultivation in Gujarat. ML-150 variety has been developed by Punjab Agriculture University (PAU), Ludhiana through intervarietal hybridization of Type 8 × Type 36. It is a medium-to-late variety with red leaf margins and yellow seeds (Pandey and Roy, 2011).

12.11 SOWING TIME

Fenugreek is usually sown in October and mid-November. The optimum sowing period is mid-October for its better plant establishment and better vegetative growth. The staggered sowing is helpful for constant supply of green fodder during January–March. The seeds are generally broadcast. The fenugreek is sown in line at 2–3-cm soil depths.

12.12 SEED TREATMENT AND INOCULATION

To treat the seed with fungicide, apply Thiram @ 3 g/kg of seed. Seeds should be inoculated with bacteria before sowing for fixation of atmospheric nitrogen. Apply 20 g rhizobium for 1 kg of seed before sowing for better nodulation.

12.13 SEED RATE AND SOWING METHOD

The seed rate is 30–36 kg/ha for the sole crop by broadcasting. The seed rate is 15–20 kg/ha for intercropping systems. For dry areas, seed rate should be 40 kg/ha as a mixed cropping system. The seed can be sown in line with seed drill or seed cum fertilizer drill machines.

12.14 SPACING

For line sowing, row-to-row spacing is maintained about 25–30 cm. but depending upon the soil, climate, and irrigation facilities, it may be increased. In dry areas, the spacing may be increased up to 50 cm when it is grown in intercropping systems with oats and barley.

12.15 CROPPING SYSTEMS

For quality and quantity forge production, fenugreek is grown in association with oats and barley. The seed rate is reduced to 60–65% of normal seed rates for both the crops. During the winter season, anjan, Rhodes grass, Setaria, Guinea grass, etc. remain dormant; fenugreek is oversown to get an extra harvest of fenugreek green fodder.

- Some of the important rotations of the fenugreek are given below:
- Maize–fenugreek
- Oats + fenugreek
- Barley + fenugreek
- Pearl millet–fenugreek
- Fenugreek–pearl millet

12.16 NUTRIENT MANAGEMENT (MANURES AND FERTILIZERS)

The crop requires less use of fertilizers. Application of organic manure, nitrogen, and phosphorus has been found to give higher green fodder yield. Farm yard manure or compost at 5–10 t/ha is desired for the crop under rainfed conditions. Apply 20–25 kg/ha N and 50–60 kg/ha P_2O_5 at the time of sowing as a basal is beneficial for this crop.

12.17 WATER MANAGEMENT

The seed requires high moisture or one irrigation immediately after sowing for quick germination. Sometimes, presowing irrigation is advisable for getting proper soil moisture germination if soil moisture is not sufficient to germinate the seed. In winter, if rain fails, then supplementary irrigation may be required for better growth.

12.18 INTERCULTURE OPERATION AND WEED MANAGEMENT

Broadcasted fenugreek fodder crop generally is not practiced intercultural operation and weeding. Crop weed competition is more common in early stage of crop growth (25–35 days after sowing). Normally in most growing areas of Southeast Asia, weeds are controlled by hand weeding. This crop often is considered a low-input crop and therefore weeding may not take place. Regular cultivated field weeding does not require.

12.19 INSECT-PEST AND DISEASE MANAGEMENT

Some important insects are stem fly: *Ophiomyia* spp., cowpea aphid: *Aphis craccivora* Koch, serpentine leaf miner: *Liriomyza trifolii* Burgess, thrips: *Scirtothrips dorsalis* Hood, lucerne weevil: *Hypera postica* Gyllenhal, spotted pod borer: *Maruca testulalis* Geyer, mite: *Tetranychus cucurbitae* Rahman and Sapra. Some important diseases are given: Cercospora leaf spot: *Cercospora traversiana* Sacc., charcoal rot: *Macrophomina phaseolina* (Tassi) Goid; powdery mildew: *Erysiphe polygoni* DC, *Leveillula taurica* (Lev.) arm; downy mildew: *Peronospora trigonellae* Gaum; rust: *Uromyces trigonellae* Pass; root rot/collar rot/foot rot: *Rhizoctonia solani* Kuhn; *Sclerotium rolfsii* Sacc.; damping off: *Pythium aphanidermatum* (Edson) Fitzp; *Fusarium* wilt: *Fusarium oxysporum* (Schlecht.); yellow mosaic disease: bean yellow mosaic virus.

12.20 HARVESTING

The suitable time for green forage harvesting is small pod development stage. Generally, fenugreek first harvesting is done at 60 days after sowing and subsequent second cut is taken in 5 weeks after first cut. Seed production fenugreek crop harvested at fully ripening stage.

12.21 YIELD

Irrigated fenugreek crop gives 250–350 q/ha green fodder but rainfed crop gives 150–250 q/ha.

12.22 SEED PRODUCTION

During seed production, extra care should be taken to avoid mixing with other crop variety and weed seeds. For obtaining pure seed, fenugreek seed production unit should maintain isolation distance to avoid cross-pollination. Seed rate of fenugreek is 20 kg/ha optimum with 45–50 cm spacing in line-sown crop. The average seed yield of fenugreek is 6–8 q/ha.

12.23 NUTRITIVE VALUE

The chemical compositions of fenugreek are crude protein (CP) (3.6%), ether extract (3.5%), crude fiber (7.1%), nitrogen-free extract (9.5%), and ash (2%) at flowering stage on dry weight basis. One cutting of fenugreek has been reported to yield between 5 and 8 dry matter ton per acre. CP content of fenugreek cut at 15 and 19 weeks and of early bloom alfalfa (19.8%, 15.7%, and 18.2%, respectively) was not different (Darby, 2004).

12.24 TOXICITIES

Because of the high content of coumarin-like compounds in fenugreek, it may interfere with the activity and dosing of anticoagulants and anti-platelet drugs and causes birth defects in animals and it can pass through the placenta (Ouzir et al., 2016).

12.25 UTILIZATION

Fenugreek is used as an herb (dried or fresh leaves) spice (seeds) and vegetable (fresh leaves and sprouts). Sotolon is the chemical responsible for fenugreek's distinctive sweet smell. Fenugreek used for both whole and powdered in the preparation of pickles, vegetable dishes, *dals*, and spice mixes such as *panch phoron* and *sambar* powder. They are often roasted to reduce bitterness and enhance flavor. The maple aroma and flavor of fenugreek have led to its use in many baked goods and imitation maple syrup. Seeds are also ground and used in curries. Young seedlings and other portions of fresh plant materials are eaten as vegetables. Fenugreek is also used as a livestock feed. As a medicinal plant, fenugreek has traditionally been considered a carminative, demulcent, expectorant, laxative, and stomachic. The plant has also been used against bronchitis, fevers, sore throats, wounds swollen glands, skin irritations, and diabetes. Fenugreek has been used to promote lactation and as an aphrodisiac. Fenugreek seeds have been used as an oral insulin substitute and seed extracts are used to lower blood glucose (Ouzir et al., 2016).

KEYWORDS

- **metha**
- **agronomic management**
- **fodder yield**
- **nutritive value**
- **toxicities**
- **utilization**

REFERENCES

Acharya, S. N.; Thomas, J. E.; Basu, S. K. Fenugreek, an Alternative Crop for Semiarid Regions of North America. *Crop Sci.* **2008,** *48*, 841–853.

Chatterjee, B. N.; Das, P. K. *Forage Crop Production Principles and Practices*; Oxford & IBH Publishing Co. Pvt. Ltd.: Kolkata, 1989.

Darby, H. Fenugreek: An Ancient Forage with a New Twist. *NODPA News* **2004,** *4* (1), 12–13.

Ouzir, M.; Bairi, K. E.; Amzazi, S. Toxicological Properties of Fenugreek *(Trigonella foenum graecum). Food Chem. Toxicol.* **2016,** *96,* 145–154.

Pandey, K. C.; Roy, A. K. *Forage Crops Varieties*; IGFRI: Jhansi, India, 2011.

Parthasarathy, V. A.; Kandinnan, K.; Srinivasan, V. *Fenugreek. Organic Spices;* New India Publishing Agencies: Delhi, 2000; p 694.

CHAPTER 13

SENJI (SWEET CLOVER)

MD. HEDAYETULLAH[1*], PARVEEN ZAMAN[2], and RAGHUNATH SADHUKHAN[3]

[1]*Department of Agronomy, Bidhan Chandra Krishi Viswavidyalaya, Mohanpur, Kalyani 741235, Nadia, West Bengal, India*

[2]*Assistant Director of Agriculture, Pulse and Oilseed Research Sub-station, Department of Agriculture, Government of West Bengal, Beldanga 742133, Murshidabad, West Bengal, India*

[3]*Department of Genetics and Plant Breeding, Bidhan Chandra Krishi Viswavidyalaya, Mohanpur 741252, Nadia, West Bengal, India*

[*]*Corresponding author. E-mail: heaye.bckv@gmail.com*

ABSTRACT

It is popularly known as senji or Indian clover or King Island clover. *Melilotus* spp. is not true clovers. Two common species are *Melilotus alba* (white sweet clover) and *Melilotus officinalis* (L.) Pall. (yellow sweet clover). It is a common leguminous weed and also a cultivated winter annual forage crop of Northern India, particularly of Punjab and Uttar Pradesh. It is persistent and aggressive in growth and does not require good tillage, fertile soil, much irrigation, and aftercare. The average yield of green fodder ranges between 300 and 400 q/ha depending upon the soil type and management practices. The average seed yield ranges between 8 and 10 q/ha. The chemical composition of green fodder of senji is crude protein 3.8%, ether extract 0.6%, crude fiber 5.8%, ash 2.3%, and nitrogen-free extract 97%.

13.1 BOTANICAL CLASSIFICATION

Kingdom: Plantae
Orders: Fabales
Family: Fabaceae
Subfamily: Faboideae
Tribe: Trifolieae
Genus: *Melilotous* L.
Species: *indica*
Binomial name: *Melilotus indica* L.

13.2 BOTANICAL NAME

Melilotus indica L. or *Melilotus officinalis* L. or *Melilotous parviflora* Desf.; synonyms: *Trifolium indicum L.*

13.3 COMMON NAME

Senji or sweet clover or melilot or annual yellow sweet clover or common melilot (Hindi: senji; English: Indian sweet clover or King Island clover).

13.4 INTRODUCTION

This legume is commonly named for its sweet smell, which is due to the presence of coumarin in its tissues. It is popularly known as senji or Indian clover or King Island clover. *Melilotus* spp. is not true clovers. Two common species are *Melilotus alba* (white sweet clover) and *Melilotus officinalis* (L.) Pall. (yellow sweet clover). It is a common leguminous weed and also a cultivated winter annual forage crop of Northern India, particularly of Punjab and Uttar Pradesh. It is persistent and aggressive in growth and does not require good tillage, fertile soil, much irrigation, and aftercare.

13.5 ORIGIN AND DISTRIBUTION

Indian clover is indigenous to India and it has been under cultivation since ancient times. Originally from Europe and Asia, now it is found worldwide. It is grown in Pakistan and to a limited extent in Argentina, Tasmania, Canada, Southern Australia, and Southern United States. Originally, it grew as a weed but when farmers found it is useful for farm animals, they started cultivated it as a fodder crops.

13.6 PLANT CHARACTERISTICS

It is an annual herb of about 80–100 cm height. The stems are hollow and cylindrical; leaves are pinnately trifoliate, stipules adnate, flowers small and yellow on slender axillary racemes, and pods small pod. The midrib is reddish in young leaves but this reddish color disappears in later stage.

13.7 CLIMATIC REQUIREMENT

Sweet clover grows well under a wide range of climatic and soil conditions. The crop thrives well in cool and mild winter along with bright sunlight in temperate and subtropical regions. High humidity is detrimental for crop growth and higher green fodder production. The Indian clover can tolerate drought for 5–6 weeks and more heat compared to other species. This crop prefers loamy and alluvial soil for its growth. It prefers neutral to slightly alkaline soil and is tolerant to saline conditions. The plant cannot survive under waterlogged situation (ICAR, 2006).

13.8 SOIL AND ITS PREPARATION

This crop prefers loamy and alluvial soil for its growth. It prefers neutral to slightly alkaline soil and is tolerant to saline conditions. The plant cannot survive under waterlogged situation. Indian clover can withstand low moisture-retention capacity of light sandy soil. In heavy soil, drainage facility should be there for draining excess water. Sweet clover can grow in light land preparation but for higher green fodder production, field should be well prepared. Before plowing, well decomposed farmyard manure or organic manure can broadcast for higher green fodder production. Generally, field is prepared by one plowing with moldboard plow followed by one or two cross harrowing. After final plowing, land is planked by wooden or iron plank to well establishment of seeds. To ensure good germination, presowing irrigation is given (Chatterjee and Das, 1989).

13.9 VARIETIES

Selections at Sirsa have led to the development of few varieties such as FS 14 and FS 18. Some improved varieties of Indian clover (Pandey and Roy, 2011) are given below:

FOS-1: Choudhury Charan Singh Hariyana Agriculture University (CCSHAU), Hisar developed the variety through selection from local material.

Senji safed-76: The variety was developed through hybridization between strain no. 341 and strain no. 174 by Punjab Agriculture University (PAU), Ludhiana.

YSL-106: The variety is a derivative of FOS-1 × strain no. 807 by PAU, Ludhiana. The variety has been recommended for cultivation under irrigated and timely sown conditions of rabi season in Punjab.

PC-5: The variety has been developed through hybridization between S 67 × Karak by PAU, Ludhiana. The variety has been recommended for cultivation under irrigated and timely sown condition of rabi in Punjab. The variety is of long duration and matures in 172 days. It provides green fodder yield of 23 t/ha.

HFWS-55: This variety is developed through selection from local material by CCSHAU, Hisar. The variety is released for all senji growing areas of Haryana. It is medium to late in maturity, leafy, palatable, and white flowered. It provides 45–50 t/ha green fodder and 8–10 dry fodder and seed yield is 12.0–15.0 q/ha.

13.10 SOWING TIME

Indian clover is usually sown in late September and mid-November. The optimum sowing time is mid-October for its better plant establishment.

Staggered sowing is helpful for constant supply of green fodder during January–March. Hard seed coat creates problem of germination of sweet clover, hence scarification is must for speedy and better germination. The scarification is done by rubbing with stones or sand on a hard or concrete floor. The seeds are usually sown by drilling or by broadcasting. The seeds are soaked overnight and sown in standing water.

13.11 SEED TREATMENT AND SEED INOCULATION

To treat the seed with fungicide apply Thiram @ 3 g/kg of seed. Seeds should be inoculated with bacteria *Azotobacter* before sowing for fixation of atmospheric nitrogen. Apply 20 g *Azotobacter* for 1 kg of seed before sowing for better nodulation. *Rhizobium* inoculation is not essential for senji crop.

13.12 SEED RATE AND SOWING METHOD

The unhooked seed rate is 50 kg/ha for the sole crop by broadcasting. For line sowing, unhusked seed rate should be 35 kg/ha and husked 25 kg/ha.

13.13 SPACING

For line sowing row-to-row spacing is maintained about 25–30 cm.

13.14 CROPPING SYSTEMS

For quality and quantity forage production, senji is grown in association with oats and barley. The seed rate is reduced to 60–65% of normal seed rates for both the crops. During the winter season hybrid Napier, blue panic, anjan, Rhodes grass, setaria, guinea grass, etc. remain dormant; senji is oversown to get an extra harvest of senji.

Some of the important rotations of the senji are given bellow:

- Maize–senji
- Lucerne + senji

- Shaftal or hubam + senji
- Pearl millet–senji

13.15 NUTRIENT MANAGEMENT (MANURES AND FERTILIZERS)

Application of organic manure, nitrogen, and phosphorus has been found to give higher green fodder yield. Farmyard manure or compost at 5–10 t/ha is desired for the crop under rainfed conditions. Apply 20–25 kg/ha N and 50–60 kg/ha P_2O_5/ha at the sowing time as a basal.

13.16 WATER MANAGEMENT

The hard coated seed requires high moisture or standing water or one irrigation immediately after sowing for quick germination. In winter, if rain fails then supplementary irrigation may be required for better growth.

13.17 INTERCULTURE OPERATION AND WEED MANAGEMENT

Broadcasted senji fodder crop generally does not practice intercultural operation and weeding. Crop–weed competition is more common in early stage of crop growth (25–35 days after sowing). Normally in most growing areas of Southeast Asia, weeds are controlled by hand weeding (Ogle et al., 2008). This crop is often considered as a low-input crop with lower returns and therefore weeding may not take place.

13.18 INSECT-PEST AND DISEASE MANAGEMENT

No serious pests and diseases have been reported in India on Indian clover. The diseases such as soft rot (*Phytophthora cactorum*), a soil borne and damping off, or seedling blight are reported in other countries. Sweet-clover weevil (*Sitona cylindricollis*) reduces sweetclover stands. Brown root rot, common leaf spot, and gray stem canker can also pose problems. “Yukon” is reported to be resistant to brown rot and gray stem canker. Control common leaf spot by cutting, before defoliation becomes severe.

Gray stem canker can be controlled with a good crop rotation and by cutting fields cleanly (Smith and Gorz, 1965).

13.19 HARVESTING

Indian clover is harvested at small pod development stage when the pods are just being formed. In sole crop, senji sown by mid-October makes two cuttings possible, first at 60 days after sowing and second 40 days after first cut. Mixture forage crop with oats and barley, and senji is harvested at 90–100 days after sowing for single cut.

13.20 YIELD

The average yield of green fodder ranges between 300 and 400 q/ha depending upon the soil type and management practices.

13.21 SEED PRODUCTION

The seed crop of Indian clover requires clean cultivation and balance fertilization. The seed rate is reduced to 75% of that fodder crops. The average seed yield ranges between 8 and 10 q/ha.

13.22 NUTRITIVE VALUE

The senji fodder crops are highly palatable and nutritious. It is rich in protein content than berseem and lucerne. The chemical composition of green fodder of senji is crude protein 3.8%, ether extract 0.6%, crude fiber 5.8%, ash 2.3%, and nitrogen-free extract 97%.

13.23 UTILIZATION

Sweet clover can be used as pasture or livestock feed. It is most palatable in spring and early summer, but livestock may need time to adjust to the bitter taste of coumarin in the plant. Prior to World War II, before the common use of commercial agricultural fertilizers, the plant was commonly used as a cover crop to increase nitrogen content and improve subsoil water capacity in poor soils. Sweet clover is a major source of nectar for domestic honey bees as hives near sweet clover can yield up to 200 lb of honey in a year (USDA, 1937). Sweet clover has been used as a phytoremediation and phytodegradation plant for treatment of soils contaminated with dioxins. In the chemical industry, dicoumarol is extracted from the plant to produce rodenticides.

KEYWORDS

- **senji**
- **agronomic management**
- **fodder yield**
- **nutritive value**

REFERENCES

Chatterjee, B. N.; Das, P. K. *Forage Crop Production Principles and Practices*; Oxford & IBH Publishing Co. Pvt. Ltd.: Kolkata, 1989.

ICAR. Handbook of Agriculture. ICAR: New Delhi, 2006.

Ogle, D.; John, L. S.; Tilley, D. *Plant Guide for Yellow Sweet Clover [Melilotus officinalis (L.)] Lam. and White Sweet Clover* (*M. alba*) *Medik.*; USDA, Natural Resources Conservation Service, Idaho Plant Materials Center: Aberdeen, 2008 (ID. 83210).

Pandey, K. C.; Roy, A. K. *Forage Crops Varieties*; IGFRI: Jhansi, India, 2011.

Smith, W. K.; Gorz, H. J. Sweet Clover Improvement. *Adv. Agron.* **1965,** *17*, 163–231.

USDA Forest Service. *Range Plant Handbook.* US Government Printing Office: Washington, DC, 1937.

CHAPTER 14

MUNG BEAN (GREEN GRAM)

KAJAL SENGUPTA*

Department of Agronomy, Bidhan Chandra Krishi Viswavidyalaya, Mohanpur, Kalyani 741235, Nadia, West Bengal, India

**E-mail: drkajalsengupta@gmail.com*

ABSTRACT

Mung bean (*Vigna radiata* L. Wilczek) is an important annual leafy legume crop grown widely in Southeast Asia. It is a short duration (60–90 days) crop that can be grown twice a year, that is, in spring and autumn. In India, mung bean is mainly cultivated as a pulse crop in three different seasons, namely, spring/summer, rainy (kharif), and winter (rabi); however, this can be used as a fodder particularly during the lean period. It is consumed as whole grains, sprouted form as well as dhal in a variety of ways in homes. It is also used as a green manuring crop. Mung bean can be used as a feed for cattle; even husk of the seed can be soaked in water and used as cattle feed. The green gram fodder contains, on an average, 10–15% crude protein, 20–26% crude fiber, 2–2.5% ether extract, 40–49% nitrogen-free extract, and 11–15% ash on dry matter basis. Fodder value of second cut is generally low.

14.1 BOTANICAL CLASSIFICATION

Kingdom: Plantae
Division: Magnoliophyta
Class: Magnoliopsida
Order: Fabales
Family: Fabaceae

Sub family: Faboideae
Genus: *Vigna*
Species: *radiata*
Binomial name: *Vigna radiata* (L.) R. Wilczek
Synonyms: *Phaseolus aureus* Roxb/*Phaseolus radiatus* L.
Diploid chromosome number of 2n = 22

14.2 BOTANICAL NAME

Vigna radiata (L.) R. Wilczek.

14.3 COMMON NAME

Green gram and mung bean. It is known as *mung* in Hindi, Oriya, Bengali, Punjabi, and Urdu language. Green gram is also known as *mudga* or *moong* in Sanskrit language.

14.4 INTRODUCTION

Mung Bean is a common pulse crop; however, this can be used as a fodder particularly during the lean period. Being a legume fodder, it has special significance because of high herbage protein and partial independence for their nitrogen needs. Mung bean, also known as mung, moong, mash bean, munggo, or monggo, green gram, golden gram, and green soy, is the seed of *Vigna radiata* which is native to India. The split bean is known as *moong dal*, which is green with the husk, and yellow when dehusked. The beans are small, ovoid in shape, and green in color. The English word "*mung*" derives from the Hindi *moong*. It is also known by the names *hesaru bele* (Kannada), *moog* (Marathi), *payiru* (Tamil), *cheru payaru* (Malayalam), and *pesalu* (Telugu). In the Philippines, it is called *munggo* or *monggo*. The mung bean is one of many species moved from the genus *Phaseolus* to *Vigna* and is still often seen cited as *Phaseolus aureus* or *Phaseolus radiatus*. These are all the same plant.

Mung bean (*Vigna radiata* L. Wilczek) is an important annual leafy legume crop grown widely in Southeast Asia. It is a short duration (60–90 days) crop that can be grown twice a year, that is, in spring and autumn.

It has less water requirement as compared to other summer crops (Abd El-Salam et al., 2013). Moreover, it is drought tolerant that can withstand adverse environmental conditions, and hence successfully be grown in rainfed areas. It has the potential to enrich soil through atmospheric nitrogen fixation, it fixes atmospheric nitrogen at 50–100 kg per hectare annually. Although mung bean is grown mostly for grain production, it can be used as a double-purpose (forage and seed) crop. Although mung bean is grown mostly for seeds production, it can be used as a dual-purpose crop (early season forage production followed by seed production). Green gram can be used as green manure, cover crop, and as fodder in cut-and-carry system and as a concentrate feed. It can be incorporated into cereal cropping systems as a legume ley to address soil fertility decline and is used as an intercrop species with maize to provide better legume/grass feed quality (Abd El-Salam et al., 2008; Abd El-Salam and El-Habbasha, 2008). Mung bean can be used as green forage for livestock and may give farmers a chance to improve the quantity and quality of forage available for clipping or grazing.

The promising multicutting mung bean varieties with the high nutritive value could effectively be employed to narrow the summer green forage gape and overcome the critical forage shortage period.

14.5 ORIGIN AND DISTRIBUTION

The mung bean is thought to have originated from the Indian subcontinent where it was domesticated as early as 1500 B.C. It was selected from a wild species of mung bean (subsp. *sublobata*), which is widely distributed in Asia and Africa. The crop is produced on a large scale in Southern and Eastern Asia. China, Australia, the United States, and other countries have started growing the crop.

Cultivated mung beans were introduced to Southern and Eastern Asia, Africa, Austronesia, the Americas, and the West Indies. It is now widespread throughout the tropics and is found from sea level up to an altitude of 1850 m in the Himalayas (Lambrides and Godwin, 2006; Mogotsi, 2006).

Mung bean production is mainly (90%) situated in Asia: India is the largest producer with more than 50% of world production but consumes almost its entire production. China produces large amounts of mung beans, which represents 19% of its legume production. Thailand is the main exporter and its production increased by 22% per year between 1980 and 2000 (Lambrides and Godwin, 2006). Though it is produced in many African countries, the mung bean is not a major crop there (Mogotsi, 2006).

14.6 PRODUCTION AREA

Green gram is widely grown in India, Pakistan, Bangladesh, Myanmar, Thailand, Philippines, China, and Indonesia. Its cultivation has now spread to Australia, Africa, and America. About 70% of world's green gram production comes from India. The Indian states producing this pulse are Madhya Pradesh, Maharashtra, Uttar Pradesh, Punjab, Andhra Pradesh, Rajasthan, Karnataka, West Bengal, and Tamil Nadu.

14.7 PLANT CHARACTERISTICS

Mung bean plant is an erect or suberect, deep rooted, and hairy annual herb. The growth habit includes both upright and vine types. Leaves are alternate, trifoliate (occasionally quadra- or pentafoliate). Inflorescence is an axillary or terminal raceme with a cluster of 10–20 papilionaceous flowers. Flowers are pale yellow in color, borne in clusters of 10–20 near the top of the plant. Pods are long, thin, and cylindrical, slightly bulged over the seeds, and the color varies from black and brown to pale gray when matured. They are 7–10-cm long, each having 8–15 seeds. They develop in clusters at a leaf axil, with typically 20–40 pods per plant. Seeds color exhibits a wide range of variations at maturity from yellow, greenish yellow, light green, and shiny green to dark green, dull green, black, brown, and green mottled with black.

Wild types tend to be prostrate while cultivated types are more erect (Lambrides and Godwin, 2006). The stems are multibranched, sometimes twining at the tips (Mogotsi, 2006). The leaves are alternate, trifoliolate with elliptical to ovate leaflets, 5–18-cm long × 3–15-cm broad. The flowers (4–30) are papilionaceous, pale yellow or greenish in color. The pods are long, cylindrical, hairy, and pending. They contain 7–20 small, ellipsoid, or cube -shaped seeds. The seeds are variable in color: they are usually green, but can also be yellow, olive, brown, purplish brown, or black, mottled and/or ridged. Seed colors and presence or absence of a rough layer are used to distinguish different types of mung bean (Lambrides and Godwin, 2006; Mogotsi, 2006). Cultivated types are generally green or golden and can be shiny or dull depending on the presence of a texture layer (Lambrides and Godwin, 2006). Golden gram, which has yellow seeds, low seed yield, and pods that shatter at maturity, is often grown for forage or green manure. Green gram has bright green seeds, is more prolific and ripens more uniformly, with a lower tendency for pods to shatter. In India, two other types of mung beans exist, one with black seeds and one with brown seeds (Mogotsi, 2006).

14.8 CLIMATIC REQUIREMENT

Mung bean is a warm season crop, requiring 60–90 days of frost-free conditions from sowing to maturity, depending on variety. The optimum temperature range for growth is between 27°C and 30°C. Seed can be sown when the minimum temperature is above 15°C. Crop plants are susceptible to water logging. So, proper drainage system is to be provided for better growth and development of the crop.

In India and Bangladesh, the crop is grown during two seasons. One is the rabi (winter) season (starting November), and the other is the summer or rainy (kharif) season (starting March). Green gram is tropical (or subtropical) crop, and requires warm temperatures (optimal at 30–35°C). High humidity and excess rainfall late in the season can result in disease problems and harvesting losses owing to delayed maturity.

The mung bean is a fast-growing, warm season legume. It reaches maturity very quickly under tropical and subtropical conditions where optimal temperatures are about 28–30°C and always above 15°C. It can be sown during summer and autumn. It does not require large amounts of water (600–1000 mm rainfall/year) and is tolerant of drought. It is sensitive to water logging. High moisture at maturity tends to spoil the seeds that may sprout before being harvested.

14.9 SOIL AND ITS PREPARATION

The mung bean grows on a wide range of soils such as red laterite soils, black cotton soils, and sandy soils but prefers well-drained loams or sandy loams, with a pH ranging from 5 to 8. It is somewhat tolerant to saline soils (Mogotsi, 2006).

A well-drained loamy to sandy loam soil is best for its cultivation. This crop prefers loamy and alluvial soil for its growth. It prefers neutral soil. Loamy soil is best for green gram cultivation. Mung beans do well on fertile, sandy loam soils with good internal drainage and a pH of between 6.5 and 7.5. The crop does not grow well on saline and alkaline soil or waterlogged soils.

Green gram requires proper drainage and ample aeration in the field so that activities of the nitrogen-fixing bacteria are not hampered at any stage of plant growth. Crop plants are susceptible to water logging. So, proper drainage system is to be provided for better growth and development of the crop.

14.10 VARIETIES

Some important varieties of mung bean grown in India are IPM 2-3, Co-6, TM 96-2, Vamban 2, Vamban 3 (for Tamil Nadu); IPM 02–14 and 2-3, HUM 1, PKVAKM-4, COGG 912, KKM 3, LGG 460, TARM-1, OBGG 52 (for Karnataka); HUM 1, BM 2002-1, PKVAKM-4, BM 4, TARM 2, Meha (for Madhya Pradesh and Maharashtra); Pant Mung 5, Pant Mung 4, Narendra Mung 1 (for Uttar Pradesh and Uttarakhand); Madhira 429, Pusa-9072, WGG-2, IPM-02-14, OUM 11-5, CoGG-912, LGG-460, LGG-450, LGG-407, and TM 96-2 (for Andhra Pradesh). Varieties found suitable under West Bengal (India) conditions are Sonali, Panna, Sukumar, Bireswar, Samrat, PDM-54, and Pant Mung 2. Most of these varieties are harvested in 40–45 days with 25–30 green pods per plant. From any variety approximate 20–25 t/ha green fodder yield and 900 kg/ha of seed yield can be obtained.

Some South African varieties are: Chainat 60, BPI Mg 7 and Merpati; varieties suitable for Bangladesh and Pakistan are BARI Mung-2 (Kanti), BARI Mung-3 (Progoti), BARI Mung-4 (Rupsha), BARI Mung-5 (Taiwani), BARI Mung-6, BINA Mung-1, BINA Mung-2, BINA Mung-3, BINA Mung-4, BINA Mung-5, BINA Mung-6, BINA Mung-7, BINA Mung-8, BU mug-1, BU mug-2, BU mug-3, and BU mug-4, NM-2006, NM-51, NM-54, NM-98, Chakwal Mung-97,

Chakwal Mung-2006; some Australian varieties are: Jade-AU, Crystal, and Celera II-AU.

14.11 SOWING TIME

For the summer or spring crop, mung bean should be sown after the harvest of last crop (potato, sugarcane, mustard, cotton, etc.). The first fortnight of March is most suitable for spring/summer cultivation. In India, kharif mung bean is generally sown during the last week of June to mid- or first week of July.

14.12 SEED TREATMENT AND INOCULATION

Well dried seed may be treated with captan or Thiram @ 3 g/kg seed against any seed borne fungal disease. *Rhizobium* inoculation is highly recommended in fields where mung bean cultivation is taken up for the first time.

14.13 SEED RATE AND SOWING METHOD

Seeds which are healthy, undamaged, and free from insect-pests and fungi should be selected. The seed rate varies with seed size and season. In the case of bold seed types, a seed rate of 40 kg/ha is appropriate in spring and autumn, and 30 kg/ha in summer. It is advised to establish a plant population of 25–30 plants/m^2 for obtaining good seed yield.

14.14 SPACING

Line sowing is more advantageous as it requires less seed, produces a more even crop that is easier to manage and will have higher yield potential. The space between the lines is kept at 25–30 cm and 10 cm between the plants. Broadcasting of seed, although an established practice, makes weeding, crop management, and harvest much more labor intensive and significantly reduces crop productivity and economic return.

27.15 CROPPING SYSTEMS

Some important crop rotations with mung bean are:

Main rotation patterns:

Hybrid pearl millet–cluster bean–mung bean–rapeseed–mustard
Hybrid pearl millet–mung bean–sunflower/safflower
Mung bean–fallow–wheat/barley
Mung bean (summer)–rice–wheat
Mung bean–rice–mustard
Mung bean–maize
Mung bean–maize + mung bean–wheat
Mung bean–maize (early)–potato (early)–wheat

Promising intercropping systems that include legumes in India:

Maize + mung bean (1:1)
Jute + mung bean (1:1)
Pearl millet + mung bean (1:1)
Cotton + mung bean (1:1)
Sugarcane + mung bean (1:2)
Sorghum + mung bean (1:1)

Mung beans are commonly grown along with sorghum for dual purposes, that is, fodders and gains.

14.16 NUTRIENT MANAGEMENT (MANURES AND FERTILIZERS)

The application of well decomposed 10 t of farmyard manure (FYM) provides the desired quality to the soil. The FYM is mixed with soil 1 month before sowing. *Trichoderma* inoculation is recommended and for that culture of *Trichoderma viride* @ 5 kg/ha with FYM is applied to soil before sowing (this mixture is kept under partial shade for 4–5 days before application). In addition to the above, a fertilizer mix containing N, P_2O_5, and K_2O at the rate of 0, 60, and 40 kg/ha, respectively, is broadcasted and incorporated into the soil before sowing. Sometimes top dressing of N at 10 kg/ha is done at preflowering stage for better fodder yield.

14.17 WATER MANAGEMENT

Usually, one light irrigation is given just before sowing or just after seedling emergence. Later, apply two to three more irrigations at 15-day intervals during the dry season. The crop may be irrigated depending upon weather, soil, and field conditions. The last irrigation should be stopped about 45–50 days after sowing. Generally, no irrigation is needed during the rainy season except when drought occurs.

14.18 INTERCULTURE OPERATION AND WEED MANAGEMENT

Weed control options are limited in mung bean and the most effective practice is to select a field with lower weed pressure. Weed control can be obtained either by manual weeding or by using herbicides, although herbicide application is not advocated.

14.19 INSECT-PEST AND DISEASE MANAGEMENT

14.19.1 INSECT-PESTS

The important pests of mung bean and their control measures are described here.

Bihar hairy caterpillar (*Spodoptera litura*): The small larvae are black whereas grown up larvae are dark green with black triangular spots on body. Its moth lays eggs in masses covered with brown hairs on the lower side of leaves. After hatching, first and second instar larvae feed gregariously and skeletonize the foliage. Besides leaves, they also damage floral buds, flowers, and pods.

Control: (1) Collect egg masses and young larvae with leaves and destroy them. (2) Spray with neem (commercial neem formulations or neem oil or neem seed kernel extract), *Bacillus thuringiensis* formulations and/or *Spodoptera litura* nuclear polyhedrosis virus, Novaluron 10 EC @ 1.5 mL/L or Acephate 75 SP @ 8 g/L or chlorpyrifos 20 EC @ 15 mL/L of water.

Whitefly (*Bemisia tabaci*): Whitefly is a vector of number of viral diseases especially mungbean yellow mosaic virus. The adults are tiny and very delicate and have white or smoke colored wings with which they flitter away from plants on little disturbance. Insects stick to the lower surface of leaves. The leaves of infested plants show yellowish discoloration.

Control: (1) Spray the crop with Neem oil at 20 mL/L or with Metasystox (oxydemeton methyl) 25 EC @ 3 mL/L of water to control whiteflies. (2) Use yellow sticky traps against whiteflies.

14.19.2 DISEASES

The important diseases affecting this crop and control measures are described below.

Seed and seedling rot: A number of fungi such as *Fusarium* sp., *Macrophomina phaseolina*, and *Rhizoctonia solani*, cause seed and seedling rot. This results in poor germination. It is a serious disease and sometimes resowing of the crop has to be done if it is not controlled well on time.

Control: (1) Treat the seeds with Thiram or captan @ 3 g/kg of seed, (2) sow fresh and clean seeds obtained from a healthy crop, and (3) adopt crop rotation.

Yellow mosaic: This disease is caused by virus, starting as small yellow specks along the veinlets and spreading over the lamina; the pods become thin and curl upward. The disease is transmitted by whitefly (*Bemisia tabaci*).

Control: (1) Spray the crop with Neem oil at 20 mL/L or with Metasystox (oxydemeton methyl) 25 EC @ 3 mL/L of water to control whiteflies. (2) Grow disease resistant varieties. (3) Use yellow sticky traps against whiteflies.

Cercospora leaf spot: *Cercospora* is recognized by the appearance of leaf spots that are circular to irregularly shape with grayish white centers and reddish brown to dark brown margins.

Control: (1) Spray Dithane Z-78 or Dithane M-45 @ 3.2 g/L of water. (2) Remove the plant debris from the field. (3) Remove all the infected plants and burn them. (4) Do not sow the seeds in the field which was affected last year by this disease.

Powdery mildew (*Erysiphe* sp./*Podospora* sp.): It occurs under cool temperature (20–26°C) and is favored by cloudy weather. A white–gray powdery mildew appears first in circular patches, but later spreads over the surface of the leaves, stems, and pods.

Control: (1) Spray neem seed kernel extract at 50 g/L or Neem oil at 20 mL/L twice at 10 days interval from initial disease appearance. (2) Spray

Eucalyptus leaf extract 10% at initiation of the disease and 10 days later. (3) Spray carbendazim @ 1 g/L or wettable sulfur @ 2.5 g/L of water.

14.20 HARVESTING

The plants are cut 10 cm above ground level. The crop may be cut at three occasions: at 40, 50, and 60 days after sowing. The suitable time for green forage harvesting is small pod development stage. *Vigna radiata* stalks, leaves, and husks constitute a significant proportion of livestock feed. It can be used for forage, silage, hay, and chicken feed.

Since mung beans are relatively high priced seeds (about twice the cost of other pulses), it is not cost effective to feed good quality seed to livestock. However, splits, cracked seed, and other material left after cleaning are often fed to cattle, substituting for part of the soybean ration. Mung bean plants have occasionally been used for beef cattle forage. In Afghanistan, the straw of mung bean and black gram is valued as a fodder for cattle and small stock, although ranking below lucerne and clover hay. In some locations, as in the Helmand valley, Persian clover (Dari: shaftal) is sometimes undersown in the mung bean crop to grow on after the mung has been harvested.

The mung bean can be grazed 6 weeks after planting and two grazings are usually obtained (FAO, 2012). It can be used to make hay, when it is cut as it begins to flower and then quickly dried for storage. It is possible to make hay without compromising seed harvest.

Mung bean seed yields are about 0.4 t/ha but yields as high as 2.5 t/ha can be reached with selected varieties in Asia (AVRDC, 2012). Mung beans can be sown alone or intercropped with other crops, such as other legumes, sugarcane, maize, sorghum, fodder grasses, or trees (Göhl, 1982). Intercropping can be done on a temporal basis: modern varieties ripen within 60–75 days and there is enough time to harvest another crop during the growing season. For instance, in monsoonal areas, it is possible to sow mung bean and harvest it before the monsoon season when rice is planted. It is also possible to grow mung bean on residual moisture after harvesting the rice (Mogotsi, 2006). Forage yields range from 0.64 t/ha of green matter under unfertilized conditions to about 1.8 t/ha with the addition of fertilizer (FAO, 2012).

14.21 YIELD

Usually mung bean gives 25–35 t/ha green fodder.

14.22 SEED PRODUCTION

The good quality mung bean should be with 98% genetic purity and 70% germination. In mung bean, seed inert material is restricted to 2% only. Besides, 1 kg good quality seed of mung bean should not contain more than 10 numbers of other crop seed and 10 numbers of weed seeds. General agronomic managements should be adopted for seed production of mung bean as already has been stated earlier. However, some steps are to be followed to get proper quality seeds.

The parent material, that is, nucleus/breeder's/foundation seed should be collected from the authentic source as approved by the Seed Certifying Agency. It is normally self-pollinated crop and cross pollination by insects is less than 1%. But an isolation distance of 3 m from other mung bean field is sufficient to maintain genetic purity. Proper plant protection measures should be adopted to avoid yellow mosaic disease. From 2 to 3 weeks after sowing rouging of mosaic affected plants should be done and it will be continued until harvest. At flowering stage, offtype plants should be removed on the basis of plant characteristics and flower color. Final roughing should also be done at maturity on the basis of pod characteristics.

The seed crop of mung bean should be harvested only on maturity and premature harvesting must be avoidable. After that standard postharvest technologies should be followed. Care must be taken in the threshing floor to avoid mixing among other varieties of mung bean or other crop seeds. The seeds obtained should be sun dried and moisture content should be 12%. The seeds may be stored in gunny bags and kept on wooden racks in a well ventilated and dry store.

14.23 NUTRITIVE VALUE

Mung bean as a forage is rich in protein content, being about 16–18% on the dry matter basis (El-Karmany et al., 2005; Chumpawadee et al., 2007; Abd El-Salam et al., 2008; Abd El-Salam and El-Habbasha, 2008). Mung bean, like other legume fodders, is highly palatable legume attracted by the livestock and even more nutritious in nature (Boe et al., 1991; Hediat-Ullah et al., 2012).

Generally forage quality of the second cut is low, possessed lower content of crude protein (CP), ash, and Total digestable nitrogen. This may due to that the first cut possessed greater assimilates especially in the pods than the second cut. Green fodder contained, on an average, 10–15% CP, 20–26% crude fiber, 2–2.5% ether extract.

14.24 TOXICITIES (ANTINUTRITIONAL FACTORS)

Many raw legumes (especially tropical varieties) and other high-protein grains contain antinutritional factors; toxic compounds such as protease inhibitors, alkaloids, and tannins, which reduce intake or the animal's ability to digest feed, reducing growth and production. These antinutritional factors have major effects on monogastrics (pigs, poultry), but have less effect on ruminants as microbial fermentation in the rumen can break down some toxic compounds such as the protease inhibitors. Tannins, trypsin inhibitors content of mung bean grains is low; thus can be fed in balanced diets without a problem.

14.25 UTILIZATION

Mung bean is mostly used as a cut fodder or grazed pasture. Fodder may be fed directly to livestock or used after conservation as fermented green matter (silage and haylage) or dried for products like hay, pellets, or cube concentrates. The husk of the seed can be soaked in water and used as a cattle feed (Khatik et al., 2007).

Several mung bean products are useful for livestock feeding (Vaidya, 2001) Mung beans, raw or processed, as well as split or weathered seeds. By-products of mung bean processing: mung bean bran (called chuni in India), which is the by-product of dehulling for making dhal, and the by-product of the manufacture of mung bean vermicelli. Mung bean is sometimes grown for fodder as hay, straw, or silage (Mogotsi, 2006). It is particularly valued as an early forage as it outcompetes other summer growing legumes such as cowpea or velvet bean in their early stages (Lambrides and Godwin, 2006).

FIGURE 14.1 **(See color insert.)** Green gram.

KEYWORDS

- **green gram**
- **agronomic management**
- **nutritive value**
- **yield**
- **toxicities**
- **utilization**

REFERENCES

AVRDC. *Mung Bean*, Asian Vegetable Research and Development Center—The World Vegetation Center: Shanhua, Taiwan, 2012.

Abd El-Salam, M. S.; El-Habbasha, S. F. Evaluation of Maize–Mung Bean Intercropping Systems at Different Sowing Dates for Forage Production. *Egypt J. Agron*. **2008,** *30* (2), 279–294.

Abd El-Salam, M. S.; Ashour, N. I.; Abd El-Ghany, H. M. Forage Production in Sole and Mixed Stands of Fodder Maize (*Zea mays* L.) and Mungbean (*Vigna radiata* L. Wilczek). *Bull. NRC Egypt* **2008,** *33* (1), 27–34.

Abd El-Salam, M. S.; El-Metwally, I. M.; Abd El-Ghany, H. M.; Hozayn, M. Potentiality of Using Mungbean as a Summer Legume Forage Crop Under Egyptian Condition. *J. Appl. Sci. Res.* **2013,** *9* (2), 1238–1243.

Boe, A.; Twidwell, E. K.; Kephart, K. D. Growth and Forage Yield of Cowpea and Mungbean in the Northern Great Plains. *Can. J. Plant Sci.* **1991,** *71* (7), 709–715.

Chumpawadee, S.; Chantiratikul, A.; Chantiratikul, P. Chemical Compositions and Nutritional Evaluation of Energy Feeds for Ruminant Using In Vitro Gas Production Technique. *Pak. J. Nutr*. **2007,** *6* (6), 607–612.

EL-Karamany, M. F.; Tawfic, M. M.; Amany, A.; Abdel-Aziz, M. A. Double Purpose (Forage and Seed) of Mung Bean Production 2—Response of Two Mung Bean Varieties to Replacement Part of Chemical Fertilizers by Organic Fertilizers. *Egypt J. Agric. Res.* **2005,** *2* (1), 257–268.

FAO. Grassland Index. *A Searchable Catalogue of Grass and Forage Legumes*; FAO: Rome, Italy, 2012.

Göhl, B. *Les Aliments du Bétail Sous Les Tropiques*, FAO, Division de Production et Santé Animale: Roma, Italy, 1982.

Hediat-Ullah, H., Khalil, I. H., Lightfoot, D. A., Nayab D., Imdadullah. Selecting Mungbean Genotypes for Fodder Production on the Basis of Degree of Indeterminacy and Biomass. *Pak. J. Bot.* **1982,** *44* (2), 697–703.

Khatik, K. L.; Vaishnava, C. S.; Gupta, L. Nutritional Evaluation of Green Gram (*Vigna radiata* L.) Straw in Sheep and Goats. *Indian J. Small Rumin.* **2007,** *13* (2), 196–198.

Lambrides, C. J.; Godwin, I. D. Mungbean. In *Genome Mapping and Molecular Breeding in Plants*; Kole, C., Ed.; 2006; Vol. 3, pp 69–90.

Mogotsi, K. K. *Vigna radiata* (L.) R. Wilczek. In *PROTA 1: Cereals and Pulses/Céréales et Légumes Secs. [CD-Rom]*; Brink, M., Belay, G., Eds.; PROTA: Wageningen, Netherlands, 2006.

Vaidya, S. V. *The Indian Feed Industry*; AGRIPPA, FAO: Rome, 2001.

CHAPTER 15

URD BEAN (BLACK GRAM)

KAJAL SENGUPTA* and MD. HEDAYETULLAH

Department of Agronomy, Bidhan Chandra Krishi Viswavidyalaya, Mohanpur, Kalyani 741235, Nadia, West Bengal, India

**Corresponding author. E-mail: drkajalsengupta@gmail.com*

ABSTRACT

Urd bean is an important annual leafy legume crop grown widely in Southeast Asia. It is a short duration (60–90 days) crop that can be grown twice a year, that is, in spring and autumn. In India, urd bean is mainly cultivated as pulse crop in three different seasons, namely, summer, rainy (kharif), and winter (rabi); however, this can be used as fodder particularly during the lean period. It is consumed as whole grains, sprouted form, as well as dhal in a variety of ways in homes. It is also used as a green manuring crop. Urd bean can be used as a feed for cattle; even husk of the seed can be soaked in water and used as cattle feed. The black gram fodder contains, on an average, 10–15% crude protein, 20–26% crude fiber, 2–2.5% ether extract, 40–49% nitrogen-free extract, and 11–15% ash on dry matter basis. Fodder value of second cut is generally low.

15.1 BOTANICAL CLASSIFICATION

Kingdom: Plantae
Division: Magnoliophyta
Class: Magnoliopsida
Order: Fabales
Family: Fabaceae
Subfamily: Faboideae

Genus: *Vigna*
Species: *mungo*
Binomial name: *Vigna mungo* (L.) Hepper
Synonyms: *Azukia mungo* (L.) Masam.; *Phaseolus hernandezii* Savi; *Phaseolus mungo* L.; *Phaseolus roxburghii* Wight & Arn.

15.2 BOTANICAL NAME

Vigna mungo (L.) Hepper.

15.3 COMMON NAME

Black gram and urd bean. It is known as *urd* in Hindi, Oriya, Bengali, Punjabi, and Urdu language. Black gram is also known as *kalai* or *urad* in Hindi language.

15.4 INTRODUCTION

Vigna mungo is known by various names across South and Southeast Asia. Its name in most languages of India derives from Proto-Dravidian *uẓ-untu*, borrowed into Sanskrit as *uḍida* (*Krishnamurti, 2003*). Urd bean is a common pulse crop; however, this can be used as fodder particularly during the lean period. Being a legume fodder it has special significance because of high herbage protein and partial independence for their nitrogen needs. Urd bean, also known as urad, kalai, and black gram, is the seed of *Vigna mungo* which is native to India. The split bean is known as *urad dal*, which is black with the husk, and white when dehusked. The beans are small, ovoid in shape, and black in color. The English word "*urd*" derives from the Hindi *urad*.

Urd bean is an important annual leafy legume crop grown widely in Southeast Asia. It is a short duration (60–90 days) crop that can be grown twice a year, that is, in spring and autumn. It has less water requirement as compared to other summer crops. Moreover, it is drought tolerant that can withstand adverse environmental conditions and hence successfully be grown in rainfed areas. It has the potential to enrich soil through atmospheric nitrogen fixation; it fixes atmospheric nitrogen at 50–100 kg per hectare annually. Although urd bean is grown mostly for grain production, it can be used as a double-purpose (forage and seed) crop. Although urd bean is grown mostly for seeds production, it can be used as a dual-purpose crop (early season forage production followed by seed production). Black gram can be used as green manure, cover crop, and as a fodder in cut-and-carry system; and as a concentrate feed. It can be incorporated into cereal cropping systems as a legume ley to address soil fertility decline and is used as an intercrop species with maize to provide better legume/ grass feed quality. Urd bean can be used as green forage for livestock and may give farmers a chance to improve the quantity and quality of forage available for clipping or grazing.

The promising multi-cutting urd bean varieties with the high nutritive value could effectively be employed to narrow the summer green forage gape and overcome the critical forage shortage period.

15.5 ORIGIN AND DISTRIBUTION

The urd bean is thought to have originated from the Indian subcontinent which is widely distributed in Asia and Africa. The crop is produced on a large scale in Southern and Eastern Asia. China, Australia, the United States, and other countries have started growing the crop.

Cultivated urd beans were introduced to southern and eastern Asia, Africa, Austronesia, the Americas, and the West Indies. It is now widespread throughout the tropics and is found from sea level up to an altitude of 1850 m in the Himalayas.

Urd bean production is mainly (90%) situated in Asia: India is the largest producer with more than 50% of world production but consumes almost its entire production. China produces large amounts of urd beans, which represents 19% of its legume production. Thailand is the main exporter and its production increased by 22% per year between 1980 and 2000.

15.6 PRODUCTION AREA

Black gram is widely grown in India, Pakistan, Bangladesh, Myanmar, Thailand, Philippines, China, and Indonesia. Its cultivation has now spread to Australia, Africa, and America. About 70% of world's black gram production comes from India. The Indian states producing this pulse are Madhya Pradesh, Maharashtra, Uttar Pradesh, Punjab, Andhra Pradesh, Rajasthan, Karnataka, West Bengal, and Tamil Nadu.

15.7 PLANT CHARACTERISTICS

Urd bean plant is an erect or suberect, deep rooted, and hairy annual herb. The growth habit includes both upright and vine types. Leaves are alternate and trifoliate. Inflorescence is an axillary or terminal raceme with a cluster of 10–20 papilionaceous flowers. Flowers are pale yellow in color, borne in clusters of 10–20 near the top of the plant. Pods are long, thin, and cylindrical, slightly bulged over the seeds, and the color varies from black and brown to pale gray when matured. They are 7–10-cm long, each having 8–15 seeds. They develop in clusters at a leaf axil, with typically 20–40 pods per plant. Seeds color exhibits a wide range of variations at maturity from blackish to dull black, brown, and black mottled with black.

Wild types tend to be prostrate while cultivated types are more erect. The stems are multibranched, sometimes twining at the tips. The leaves are alternate, trifoliolate with elliptical to ovate leaflets, 5–18-cm long × 3–15-cm broad. The flowers (4–30) are papilionaceous, pale yellow, or blackish in color. The pods are long, cylindrical, hairy, and pending. They contain 7–20 small, ellipsoid, or cube-shaped seeds. The seeds are variable in color: they are usually black colored seeds and presence or absence of a rough layer is used to distinguish different types of urd bean.

15.8 CLIMATIC REQUIREMENT

Urd bean is a warm season crop, requiring 60–90 days of frost-free conditions from sowing to maturity, depending on variety. The optimum temperature range for growth is between 27°C and 30°C. Seed can be sown when the minimum temperature is above 15°C. Crop plants are susceptible to water logging. So, proper drainage system is to be provided for better growth and development of the crop.

In India and Bangladesh, the crop is grown during two seasons. One is the rabi (winter) season (starting November), and the other is the summer or rainy (kharif) season (starting March). Black gram is tropical (or subtropical) crop, and requires warm temperatures (optimal at 30–35°C). High humidity and excess rainfall late in the season can result in disease problems and harvesting losses owing to delayed maturity.

The urd bean is a fast-growing, warm season legume. It reaches maturity very quickly under tropical and subtropical conditions where optimal temperatures are about 28–30°C and always above 15°C. It can be sown during summer and autumn. It does not require large amounts of water (600–1000 mm rainfall/year) and is tolerant of drought. It is sensitive to water logging. High moisture at maturity tends to spoil the seeds that may sprout before being harvested.

15.9 SOIL AND ITS PREPARATION

The urd bean grows on a wide range of soils such as red laterite soils, black cotton soils, and sandy soils but prefers well-drained loams or sandy loams, with a pH ranging from 5 to 8. It is somewhat tolerant to saline soils.

A well-drained loamy to sandy loam soil is best for its cultivation. This crop prefers loamy and alluvial soil for its growth. It prefers neutral soil. Loamy soil is best for black gram cultivation. Urd beans do well on fertile, sandy loam soils with good internal drainage and a pH of between 6.5 and 7.5. The crop does not grow well on saline and alkaline soil or waterlogged soils.

Urd bean requires proper drainage and ample aeration in the field so that activities of the nitrogen fixing bacteria are not hampered at any stage of plant growth. Crop plants are susceptible to water logging. So, proper drainage system is to be provided for better growth and development of the crop.

15.10 VARIETIES

Some important varieties of urd bean grown in India are Pant U 19, T-9, Azad Urd 1, PDU-1, KU-300, Kalindi, Naveen, Pusa-1, TAU-1, BDU-1, Pant U-35, TPU-4, Krishna, Pant U 30, Uttara, LBG-20, WBG-26, WBU-109, WBU-108, Mash-479, PU-40, Sekhar-1, Sekhar-2, Sekhar-3, LBG-752, Mash-114, Vamban-2, etc.

Some popular varieties of southern India are CO-1, CO-2, CO-3, ADT-2, ADT-3, ADT-4, ADT-5, TMV-1, VBN-1, VBN-2, VBN-3 (for rice fallows), VBN-4, KM-1, KM-2, etc.

15.11 SOWING TIME

For the summer or spring crop, urd bean should be sown after the harvest of last crop (potato, sugarcane, mustard, cotton, etc.). The first fortnight of March is most suitable for summer cultivation. In India, kharif urd bean generally sown during the last week of June to mid- or first week of July.

15.12 SEED TREATMENT AND INOCULATION

Well dried seed may be treated with captan or Thiram @ 3 g/kg seed against any seed borne fungal disease. *Rhizobium* inoculation is highly recommended in fields where urd bean cultivation is taken up for the first time.

15.13 SEED RATE AND SOWING METHOD

The seed rate of black gram is almost double for fodder production than grain seed production. Seeds which are healthy, undamaged, and free from insect-pests and fungi should be selected. The seed rate varies with seed size and season. In the case of bold seed types a seed rate of 40 kg/ha is appropriate in spring and autumn, and 30 kg/ha in summer. It is advised to establish a plant population of 25–30 plants/m^2 for obtaining good seed yield.

15.14 SPACING

For fodder production, broadcasting is generally done but line sowing is adventitious. Line sowing is more advantageous as it requires less seeds, produces a more even crop that is easier to manage, and will have higher yield potential. The space between the lines is kept at 25–30 cm and 10 cm between the plants. Broadcasting of seed, although an established practice, makes weeding, crop management and harvest much more labor intensive and significantly reduced crop productivity and economic return.

27.15 CROPPING SYSTEMS

Some important crop rotations with urd bean are:

Main rotation patterns:

Hybrid pearl millet–cluster bean–urd bean–rapeseed–mustard
Hybrid pearl millet–urd bean–sunflower/safflower
Urd bean–fallow–wheat/barley
Urd bean (summer)–rice–wheat
Urd bean–rice–mustard
Urd bean–maize
Urd bean–maize + urd bean–wheat
Urd bean–maize (early)–potato (early)–wheat

Promising intercropping systems that include legumes in India:

Maize + urd bean (1:1)
Jute + urd bean (1:1)
Pearl millet + urd bean (1:1)
Cotton + urd bean (1:1)
Sugarcane + urd bean (1:2)
Sorghum + urd bean (1:1)

The important rotations with black gram in North India (Ahlawat et al., 2000) are as given below:

Maize–wheat–urd bean
Maize–toria–urd bean
Paddy–wheat–urd bean
Black gram–wheat–urd bean
Maize–potato–urd bean

Urd beans are commonly grown along with sorghum for dual purposes, that is, fodders and gains.

15.16 NUTRIENT MANAGEMENT (MANURES AND FERTILIZERS)

The application of well decomposed 10 t of farmyard manure (FYM) provides the desired quality to the soil. The FYM is mixed with soil 1 month before sowing. *Trichoderma* inoculation is recommended and for that culture of *Trichoderma viride* @ 5 kg/ha with FYM is applied to soil before sowing (this mixture is kept under partial shade for 4–5 days before application). In addition to the above, a fertilizer mix containing N, P_2O_5, and K_2O at the rate of 0, 60, and 40 kg/ha, respectively is broadcasted and incorporated into the soil before sowing. Sometimes top dressing of N at 10 kg/ha is done at preflowering stage for better fodder yield.

15.17 WATER MANAGEMENT

Usually one light irrigation is given just before sowing or just after seedling emergence. Later, apply two to three more irrigations at 15-day intervals during the dry season. The crop may be irrigated depending upon weather, soil, and field conditions. The last irrigation should be stopped about 45–50 days after sowing. Generally, no irrigation is needed during the rainy season except when drought occurs.

15.18 INTERCULTURE OPERATION AND WEED MANAGEMENT

Weed control options are limited in urd bean and the most effective practice is to select a field with lower weed pressure. Weed control can be obtained either by manual weeding or by using herbicides, although herbicide application is not advocated.

15.19 INSECT-PEST AND DISEASE MANAGEMENT

15.19.1 INSECT-PESTS

The important pests of urd bean and their control measures are described here.

Bihar hairy caterpillar (*Spodoptera litura*): The small larvae are black whereas grown up larvae are dark black with black triangular spots on body. Its moth lays eggs in masses covered with brown hairs on the lower side of leaves. After hatching, first and second instar larvae feed gregariously and skeletonize the foliage. Besides leaves, they also damage floral buds, flowers, and pods.

Control: (1) Collect egg masses and young larvae with leaves and destroy them. (2) Spray with neem (commercial neem formulations or Neem oil or neem seed kernel extract), *Bacillus thuringiensis* formulations and/or *Spodoptera litura* nuclear polyhedrosis virus, novaluron 10 EC @ 1.5 mL/L or Acephate 75 SP @ 8 g/L or chlorpyrifos 20 EC @ 15 mL/L of water.

Whitefly (*Bemisia tabaci*): Whitefly is a vector of number of viral diseases especially urd bean yellow mosaic virus. The adults are tiny and very delicate and have white or smoke colored wings with which they flitter away from plants on little disturbance. Insects stick to the lower surface of leaves. The leaves of infested plants show yellowish discoloration.

Control: (1) Spray the crop with Neem oil at 20 mL/L or with Metasystox (oxydemeton methyl) 25 EC @ 3 mL/L of water to control whiteflies. (2) Use yellow sticky traps against whiteflies.

15.19.2 DISEASES

The important diseases affecting this crop and control measures are described below.

Seed and seedling rot: A number of fungi such as *Fusarium* sp., *Macrophomina phaseolina*, *Rhizoctonia solani*, cause seed and seedling rot. This results in poor germination. It is a serious disease and sometimes resowing of the crop has to be done if it is not controlled well on time.

Control: (1) Treat the seeds with Thiram or captan @ 3 g/kg of seed, (2) sow fresh and clean seeds obtained from a healthy crop, and (3) adopt crop rotation.

Yellow mosaic: This disease is caused by virus, starting as small yellow specks along the veinlets and spreading over the lamina; the pods become thin and curl upward. The disease is transmitted by whitefly (*Bemisia tabaci*).

Control: (1) Spray the crop with neem oil at 20 mL/L or with Metasystox (oxydemeton methyl) 25 EC @ 3 mL/L of water to control whiteflies. (2) Grow disease resistant varieties. (3) Use yellow sticky traps against whiteflies.

Cercospora leaf spot: Cercospora is recognized by the appearance of leaf spots that are circular to irregularly shaped with grayish white centers and reddish brown to dark brown margins.

Control: (1) Spray Dithane Z-78 or Dithane M-45 @ 3.2 g/L of water. (2) Remove the plant debris from the field. (3) Remove all the infected plants and burn them. (4) Do not sow the seeds in the field which was affected last year by this disease.

Powdery mildew (*Erysiphe* sp./*Podospora* sp.): It occurs under cool temperature (20–26°C) and is favored by cloudy weather. A white–gray powdery mildew appears first in circular patches, but later spreads over the surface of the leaves, stems, and pods.

Control: (1) Spray neem seed kernel extract at 50g/L or Neem oil at 20 mL/L twice at 10-days interval from initial disease appearance. (2) Spray *Eucalyptus* leaf extract 10% at initiation of the disease and 10 days later. (3) Spray carbendazim @ 1 g/L or wettable sulfur @ 2.5 g/L of water.

15.20 HARVESTING

The plants are cut 10 cm above ground level. The crop may be cut at three occasions: at 40, 50, and 60 days after sowing. The suitable time for black forage harvesting is small pod development stage. *Vigna mungo* stalks, leaves, and husks constitute a significant proportion of livestock feed. It can be used for forage, silage, hay, and chicken feed.

Since urd beans are a relatively high priced seeds (about twice the cost of other pulses), it is not cost effective to feed good quality seed to livestock. However, splits, cracked seed, and other material left after cleaning are often fed to cattle, substituting for part of the soybean ration. Urd bean plants have occasionally been used for beef cattle forage. In Afghanistan, the straw of urd bean and green gram is valued as a fodder for cattle and small stock, although ranking below lucerne and clover hay.

Urd beans can be sown alone or intercropped with other crops, such as other legumes, sugarcane, maize, sorghum, fodder grasses, or trees. Intercropping can be done on a temporal basis: modern varieties ripen within 60–75 days and there is enough time to harvest another crop during the growing season. For instance, in monsoonal areas, it is possible to sow urd bean and harvest it before the monsoon season when rice is planted. It is also possible to grow urd bean on residual moisture after harvesting the rice.

15.21 YIELD

Usually urd bean gives 25–35 t/ha green fodder.

15.22 SEED PRODUCTION

The good quality urd bean should be with 98% genetic purity and 70% germination. In urd bean, seed inert material is restricted to 2% only. Besides, 1 kg good quality seed of urd bean should not contain more than 10 numbers of other crop seed and 10 numbers of weed seeds. General agronomic managements should be adopted for seed production of urd bean as already has been stated earlier. However, some steps are to be followed to get proper quality seeds.

The parent material, that is, nucleus/breeder's/foundation seed should be collected from the authentic source as approved by the Seed Certifying Agency. It is normally self-pollinated crop and cross pollination by insects is less than 1%. But an isolation distance of 3 m from other urd bean field is sufficient to maintain genetic purity. Proper plant protection measures should be adopted to avoid yellow mosaic disease. From 2 to 3 weeks after sowing rouging of mosaic affected plants should be done and it will be continued until harvest. At flowering stage, offtype plants should be removed on the basis of plant characteristics and flower color. Final roughing should also be done at maturity on the basis of pod characteristics.

The seed crop of urd bean should be harvested only on maturity and premature harvesting must be avoidable. After that standard postharvest technologies should be followed. Care must be taken in the threshing floor to avoid mixing among other varieties of urd bean or other crop seeds. The seeds obtained should be sun dried and moisture content should be 12%. The seeds may be stored in gunny bags and kept on wooden racks in a well ventilated and dry store.

15.23 NUTRITIVE VALUE

Urd bean as a forage is rich in protein content, being about 16–18% on the dry matter basis. Urd bean, like other legume fodders is highly palatable legume attracted by the livestock and even more nutritious in nature. Generally, forage quality of the second cut is low, possessed lower content of crude protein (CP), ash, and TDN. This may due to that the first cut possessed greater assimilates especially in the pods than the second cut. Black fodder contained, on an average, 10–15% CP, 20–26 % crude fiber, and 2–2.5% ether extract. Black gram complements the essential amino acids provided in most cereals and plays an important role in the diets of the people of Nepal and India. Black gram has been shown to be useful in mitigating elevated cholesterol levels (Menon and Kurup, 1976).

15.24 TOXICITIES (ANTINUTRITIONAL FACTORS)

Many raw legumes (especially tropical varieties) and other high protein grains contain antinutritional factors; toxic compounds such as protease inhibitors, alkaloids, and tannins, which reduce intake or the animal's ability to digest feed, reducing growth, and production. These antinutritional factors have major effects on monogastrics (pigs and poultry), but have less effect on ruminants as microbial fermentation in the rumen can break down some toxic compounds such as the protease inhibitors. Tannins, trypsin inhibitors content of urd bean grains is low, thus can be fed in balanced diets without a problem. Antinutritional factors such as phenolic compounds, tannins, saponins, phytic acid, trypsin inhibitors, and enzymes related to them such as acid- and alkaline phosphatases, α-galactosidase are found in some cultivars of black gram (Suneja et al., 2011).

15.25 UTILIZATION

Urd bean is mostly used as cut fodder or grazed pasture. Fodder may be fed directly to livestock or used after conservation as fermented green matter (silage and haylage) or dried for products such as hay, pellets, or cube concentrates. The husk of the seed can be soaked in water and used as a cattle feed.

In South India, the husked dal ground into a fine paste and allowed to ferment and mixed with equal quantity of rice flour to make *dosa* and *idli*. Urd dal is also used in preparation of halva and imarti. It is also used as a green manuring crop (Ahlawat et al., 2000). Urd bean is sometimes grown for fodder as hay, straw, or silage. It is particularly valued as an early forage as it outcompetes other summer growing legumes such as cowpea or velvet bean in their early stages.

FIGURE 15.1 **(See color insert.)** Black gram.

KEYWORDS

- **black gram**
- **agronomic management**
- **nutritive value**
- **yield**
- **toxicities**
- **utilization**

REFERENCES

Ahlawat, I. P. S.; Prakash, O.; Saini, G. S. *Scientific Crop Production in India*; Aman Publishing House: Meerut, 2000.

Krishnamurti, B. *The Dravidian Languages;* Cambridge University Press: Cambridge, 2003; p 16. ISBN 978-0-521-02512-6.

Menon, P. V.; Kurup, P. A. Dietary Fibre and Cholesterol Metabolism: Effect of Fibre Rich Polysaccharide from Blackgram (*Phaseolus mungo*) on Cholesterol Metabolism in Rats Fed Normal and Atherogenic Diet. Biomedicine **1976,** *24* (4), 248–253.

Suneja, Y.; Kaur, N.; Kaur, S.; Gupta, A. K.; Kaur, N. Levels of Nutritional Constituents and Antinutritional Factors in Black Gram (*Vigna mungo* L. Hepper). *Food Res. Int.* **2011,** *44* (2), 621–628.

PART III

Nonleguminous and Nongraminaceous Forages

CHAPTER 16

SUNFLOWER (SUJYOMUKHI)

A. ZAMAN[1*] and PARVEEN ZAMAN[2]

[1]*Department of Agronomy, M. S. Swaminathan School of Agriculture, Centurion University of Technology and Management, Paralakhemundi 761211, Odisha, India*

[2]*Assistant Director of Agriculture, Pulse and Oilseed Research Sub-station, Beldanga 742133, Murshidabad, West Bengal, India*

[*]*Corresponding author. E-mail: profazaman@gmail.com*

ABSTRACT

Sunflower is used as bird food, as livestock forage, and in some industrial applications. Thc plant was first domesticated in the America. *Helianthus annuus* (wild) is a widely branched annual plant with many flower heads. The domestic sunflower, however, often possesses only a single large inflorescence (flower head) atop an unbranched stem. The name *sunflower* may have been derived from the shape of flower head, which resembles the sun, or from the impression that the blooming plant appears to slowly turn its flower toward the sun as the latter moves across the sky on a daily basis. Sunflowers can be processed into a peanut butter alternative, sunflower butter. It is also sold as food for birds and can be used directly in cooking and salads. American Indians had multiple uses for sunflowers in the past such as in bread, medical ointments, dyes, and body paints. Sunflower is a fast-growing crop with high forage yield capacity. In Cuba, fresh matter yields are 450–750 q/ha in 60–70 days in dry conditions, and up to 900 q/ha in Brazil. The local cultivar gave highest seed production up to 15 q/ha.

16.1 BOTANICAL CLASSIFICATION

Kingdom: Plantae
Orders: Asterales
Family: *Asteraceae*
Subfamily: Asteroideae
Genus: *Helianthus*
Species: *annuus* L.
Binomial name: *Helianthus annuus* L.

16.2 BOTANICAL NAME

Helianthus annuus L.; synonyms—Harpalium (Cass.).

16.3 COMMON NAME

Sunflower, sujyomukhi, flower crop.

16.4 INTRODUCTION

Helianthus annuus L., commonly known as sunflower, is large number of the genus *Helianthus* grown as an oilseed crop for its edible oil and edible fruits. This sunflower species is also used as bird food, as livestock forage, and in some industrial applications. The plant was first domesticated in the America. *H. annuus* (wilds) is a widely branched annual plant with many flower heads. The domestic sunflower, however, often possesses only a single large inflorescence (flower head) atop an unbranched stem. The name *sunflower* may have been derived from the shape of flower head, which resembles the sun, or from the impression that the blooming plant appears to slowly turn its flower toward the sun as the latter moves across the sky on a daily basis. Sunflowers can be processed into a peanut butter alternative, sunflower butter. In Germany, it is mixed with rye flour to make *Sonnen* (literally: sunflower whole seed bread), which is quite popular in German-speaking Europe. It is also sold as food for birds and can be used directly in cooking and salads. American Indians had multiple uses for sunflowers in the past such as in bread, medical ointments, dyes, and body paints.

16.5 DESCRIPTION

The plant has an erect, rough, and hairy stem, reaching typical average height of 3 m (9.8 ft). The tallest sunflower recorded is 9.17 m (Annon, 2016). Sunflower leaves are broad, coarsely toothed, rough, and mostly alternate. What is often called the "flower" of the sunflower is actually a "flower head" or pseudanthium of numerous small individual five-petaled flowers ("florets"). The outer flowers, which resemble petals, are called ray flower. Each "petal" consists of a ligule composed of fused petals of an asymmetrical ray flower. They are sexually sterile and may be yellow, red, orange, or other colors. The flowers in the center of the head are called disk flowers. These mature into fruit (sunflower "seeds"). The disk flowers are arranged spirally. Generally, each floret is oriented toward the next by approximately the golden angle, 137.5°, producing a pattern of interconnecting spirals, where the number of left spirals and the number of right spirals are successive Fibonacci numbers. Typically, there are 34 spirals in one direction and 55 in the other; however, in a very large sunflower head there could be 89 in one direction and 164 in the other (Adam, 2003). This pattern produces the most efficient packing of seeds mathematically possible within the flower head (Motloch, 2000).

Most cultivars of sunflower are variants of *H. annuus*, but four other species (all perennials) are also domesticated. This includes *Helianthus tuberosus* which produces edible tubers.

16.6 ORIGIN AND DISTRIBUTION

Sunflower seeds were brought to Europe from the Americas in the 16th century, where, along with sunflower oil, they became a widespread cooking ingredient. Although it was commonly accepted that the sunflower was first domesticated in what is now the Southeastern United States, roughly 5000 years ago, there is evidence that it was first domesticated in Mexico around 2600 B.C. These crops were found in Tabasco, Mexico at the San Andres dig site. The earliest known examples in the United States of a fully domesticated sunflower have been found in Tennessee, and date to around 2300 B.C. Many Americans used the sunflower as the symbol of their solar deity, including the races such as Aztecs and the Otomi of Mexico and the Incas in South America. In 1510, early Spanish explorers encountered the sunflower in the Americas and carried its seeds back to Europe.

The use of sunflower oil became very popular in Russia, particularly with members of the Russian Orthodox Church, because sunflower oil was one of the few oils that were allowed during Lent, according to some fasting traditions during the 18th century (Penichet et al., 2008).

16.7 PLANT CHARACTERISTICS

The flowers have a wide central disk surrounded by shorter petals in shades of cream, yellow, rust, and burgundy. Sunflower's sturdy stems can grow 10 ft or more and may hold single flowers or be multibranched. The seeds are edible and are favored by birds. They are also used to make oil and as livestock feed. There is a reason they are called sunflowers. The flower heads follow the sun. So, give some thought to where you plant your sunflowers. If you plant them on an east/west axis, you will be looking at the back of the flower heads for most of the day. Traditionally, sunflowers were a sunny yellow colored with a darker central disk. However, now we have a choice of rich chocolate browns, deep burgundies, and luscious multicolored flowers. Flowers should begin to mature in early fall. The heads will turn downward and the florets in the center disk will shrivel. The only sure way to tell if the seeds are ready to harvest is to pull a few out and open them. If they are full, they are ready. To harvest, cut the whole flower head with about 1 ft of stem attached and hang in a warm, dry, ventilated spot, away from insects and rodents. Cover the seed heads with cheesecloth or a paper bag, to catch loose seeds. Poke some small holes in the paper bag for ventilation. When the seed is completely dried and ready for use, it can be easily rubbed off the flower head and collected.

16.8 CLIMATIC REQUIREMENT

To grow best, sunflowers need full sun. They grow best in fertile, moist, well-drained soil with heavy mulch. In commercial planting, seeds are planted 45 cm (1.48 ft) apart and 2.5-cm (0.98 in.) deep. Sunflower "whole seed" (fruit) are sold as a snack food, raw, or after roasting in ovens, with or without salt and/or seasonings added.

16.9 SOIL AND ITS PREPARATION

The sunflower grows wide, shallow roots that rot and fail in standing water. Do not choose a site that puddles regularly or that tends to mud. Sunflowers do not like to have their feet wet. Sunflowers grow in poor to fertile soil as long as the gardener makes the right amendments. Mix organic compost into the top 8 in. of the soil to start, to provide the loose foundation, sunflowers require for drainage. Sunflowers grow best in locations with direct sun (6–8 h per day); they prefer long, hot summers to flower well. Sunflowers prefer a well-drained location, and prepare soil by digging an area of about 2–3 ft in circumference to a depth of about 2 ft. Though they are not too fussy, sunflowers thrive in slightly acidic to somewhat alkaline pH (6.0–7.5). Sunflowers are heavy feeders, so the soil needs to be nutrient rich with organic matter or composted (aged) manure. Or, work in a slow release granular fertilizer 8-in. deep into your soil. The field should be selected in which sunflower was not grown in the previous year unless they were of the same variety and were of equivalent or higher class and were certified. In addition, the selected field should be well drained and the soil deep, fertile, and with neutral pH.

16.10 VARIETIES

Choice of suitable variety is very important for higher yield. Some sunflower hybrids are suitable for higher green forage production; these are PAC 36 (late sown condition) >105 days, Jwalamukhi, Sungene-85, APSH-1, MSFH-8, KBSH-1, KBSH-44, PAC-1091, Pro. Sun.-09, DSH-1, etc. High-green fodder production sunflower varieties are Morden, DRSF-108, Surya, SS-56, LS-11, CO-1, CO-2, TNAUSUF-7, etc.

16.11 SOWING TIME

In preparing a bed, dig down 2 ft in depth and about 3 ft across, to ensure the soil is not too compact as sunflowers have long taproots which need to stretch out; so, the plants prefer well-dug, loose, well-draining soil. Sunflower, unlike most other crops is not season bound. Barring the periods of extreme freezing temperatures, the sowing time can be adjusted as per availability of land for planting. However, sowing should be so adjusted that the maturity of the crop does not coincide with the rains, since rains during maturity period adversely affect the seed quality. It is best to sow sunflower seeds directly into the soil after the danger of spring frost is past. Ideally, the soil temperature has reached 55–60°F.

Plants should be allowed plenty of room, especially for low-growing varieties that will branch out making rows about 30 in. apart. Plant the large seeds no more than 1-in. deep and about 6 in. apart after it has thoroughly warmed, from mid-November to late November. Two to three seeds could be planted and thin them to the strongest contenders when the plants are 6-in. tall. A light application of fertilizer mixed in at planting time will encourage strong root growth to protect them from blowing over in the wind. Experiment with plantings staggered over 5–6 weeks to keep enjoying continuous blooms.

16.12 SEED TREATMENT

The treated seeds (control, KNO_3, and hydropriming) of sunflower (*Helianthus annuus* L.) cultivar Sanbro were evaluated at germination and seedling growth for tolerance to salt (NaCl) and drought conditions induced by PEG-6000 at the same water potentials of 0.0, −0.3, −0.6, −0.9, and −1.2 MPa. Electrical conductivity values of the NaCl solutions were 0.0, 6.5, 12.7, 18.4, and 23.5 dS/m, respectively. The factors responsible for germination and early seedling growth due to salt toxicity or osmotic effect and to optimize the best hydropriming treatment for these stress conditions could easily be determined.

16.13 SEED RATE AND SOWING METHOD

The seed rate found to be optimum was 8–10 kg per hectare. The crop should be sown in rows. The depth of seedling should be 2–4 cm, having row-to-row 60 cm and plant-to-plant 30-cm distance. There was a significant interaction between seeding rate and nitrogen rate for sunflower head width, lodging, and bird damage in this experiment. These interactions indicate that sunflower seeding rate responds differently across nitrogen rates for these measurements. However, the interaction between seeding rate and nitrogen rate was difficult to interpret from a biological perspective. More data would need to be collected to help further identify if this interaction is agriculturally meaningful to farmers and sunflower production (Putt, 1997) It does appear that at low N rates and seeding rates, there is less bird damage, less lodging, and larger head widths. As N rate and seeding rates increase, the data become less clear and it is difficult to interpret the impact that these rates have on sunflower growth characteristics.

16.14 INTERCROPPING

Sunflower could be intercropped with wider ranges of winter legumes: in between rows of sunflower a short duration legume crop may be accommodated without any adverse effect on seed yield of sunflower.

16.15 CROPPING SYSTEMS

Sunflower could easily be sown as mixed crop with several winter crops particularly crop having same duration during winter months. Some important sequence cropping and intercropping are given below.

i. Sorghum–sunflower
ii. Maize–sunflower
iii. Soybean–sunflower
iv. Maize–sunflower–groundnut
v. Fallow–sunflower
vi. Sorghum–soybean–sunflower
vii. Sorghum–sunflower–maize + cowpea
viii. Groundnut–sunflower–maize + cowpea
ix. Sorghum + pigeon pea–sunflower

16.16 CROP SEQUENCE

Growing rice, nowadays can only ensure livelihood security where the resource-poor farmers are operating, if the attitudes of the rice grower remain unaltered. Sunflower could easily be accommodated during the post rice period.

Upland: Rice/maize/jowar/bajra/ragi/arhar/niger/mesta/groundnut/vegetables
Intercropping: Groundnut + sunflower/soybean + sunflower
Medium land (coastal): Rice–sunflower
Lowland (coastal): Jute–Aman paddy–sunflower

16.17 NUTRIENTS MANAGEMENT

Application of farmyard manure @ 10 t/ha, N:P_2O_5:K_2O @ 60:40:40 kg/ha as urea, single superphosphate, and muriate of potash. Full P_2O_5 and K_2O along with ½ N was applied as basal and remaining ½ N was top dressed during earthing up was recommended for the crop. However, the fertilizer required for raising a good sunflower crop is 80 kg nitrogen, 40 kg phosphorus, and 40 kg potash per hectare. At the time of planting, 50 kg nitrogen and the full amounts of phosphorus and potash should be applied as a basal dose and the remaining 30 kg nitrogen at the time of earthing, that is, after 40–45 days of crop growth.

Soil temperature should be a minimum of 7°C for planting and around 10°C for germination. Soils are often found to be deficient in nitrogen, phosphorus, and sulfur. Potassium, calcium, and magnesium are also frequently deficient in high-rainfall areas. Boron may also be required in some soils. Because the majority of sunflowers in the United States are grown in the Great Plains, recommendations for plant populations and fertilization rates are limited to this specific region and climate. Due to the temperate climate of the northeast, it is likely that optimal seeding rates and nitrogen (N) rates for sunflower production will differ from the Great Plains. A crop's N requirements are often linked to population; this study attempts to evaluate the impact of both seeding rates and N rates on sunflower yield and quality.

16.18 WATER MANAGEMENT

Sunflower seeds were sown in small furrows made by small country plow design. Initial soil moisture was 17.56% (gravimetric) in 15-cm topsoil. A light irrigation (2 cm) with help of *thali* was given for easy and even emergence of the crop. So presowing irrigation is necessary in the spring to summer seasons, and desirable for *rabi* sowing for uniform germination and better stand. Sunflower is comparatively drought tolerant and yields higher than oilseeds crops under moisture stress conditions. In rabi and zaid planting two and four irrigations, respectively, are necessary for higher yields. In kharif, if rainfall distribution is favorable, one irrigation is sufficient, to be applied between the flowering and grain filling stages. Sunflower, an oilseed crop was very sensitive to irrigation management. The highest seed yield (52.28 q/ha) was obtained with highest soil moisture regime irrespective of dates of sowing. The wetted moisture regime created increase in yield by 7.51% over drier moisture regime. The yield reduction was observed in accordance to deferred dates of sowing. Depth of irrigation considered to be optimum was 5 cm.

16.19 WEED MANAGEMENT

The sunflower plants may root lodge because of large heavy heads. Earthling, preferably before and, if needed, after irrigation around 48 days after sowing is highly desirable, 10–15-cm high earth is sufficient. One to two weeding regimes during the first 6 weeks after germination are necessary. Thereafter, growth rate is high and the crop covers the ground and smothers most of the weeds. Chemicals such as trifluralin, 6.25 pints/ha, may also be applied to control the weeds.

16.20 PLANT PROTECTION (PESTS AND DISEASES)

One of the major threats that sunflowers face today is *Fusarium*, a filamentous fungi that is found largely in soil and plants. It is a pathogen that over the years has caused an increasing amount of damage and loss of sunflower crops, some as extensive as 80% of damaged crops Downy mildew is another disease to which sunflowers are susceptible. Its susceptibility to downy mildew is particularly high due to the sunflower's way of growth and development. Sunflower seeds are generally planted only an inch deep in the ground. When such shallow planting is done in moist and soaked earth or soil, it increases the chances of diseases such as downy mildew. Another major threat to sunflower crops is broomrape, a parasite that attacks the root of the sunflower and causes extensive damage to sunflower crops, as high as 100%.

One of the major problematic weeds in the sunflower crop is wild oat, this can cause severe yield loss and should be treated if the threshold level is reached. Other potentially damaging problems include birds, rabbits, deer, mice, flooding, and frost.

Birds and squirrel will show interest in the seeds. If you plan to use the seeds, deter critters with barrier devices. As seed heads mature and flowers droop, you can cover each one with white polyspun garden fleece. If you have deer, keep them at bay with a tall wire barrier. Sunflowers are relatively insect free. A small gray moth sometimes lays its eggs in the blossoms. Pick the worms from the plants. Downy mildew, rust, and powdery mildew can also affect the plants. If fungal diseases are spotted early, spray with a general garden fungicide.

Sunflower moth: lays its eggs on the plant and the larvae feed on the flower heads, tunneling and leaving holes in the seed. Aphids and whiteflies can also be pests of the crop. Fungal diseases: Sclerotina (white mold), downy mildew, and rust—provide adequate air circulation.

The best control of diseases is prevention, by changing where you plant each year and disposing of any infected plants. Flowers in the vegetable garden are great for attracting more pollinators. To further foil squirrels, plant a coarse leaved vegetable such as squash, at their base.

The main insect problems in sunflower crops include cutworms, sunflower bud moth, sunflower stem weevil, sunflower root weevil, and the sunflower midge. Lower temperatures will increase the susceptibility of seedlings to diseases such as downy mildew.

16.21 HARVESTING

Animals can graze sunflower but waste a lot of forage when they enter the field. For that reason, ensiling is generally preferable. The best harvest time for ensiling sunflower is highly variable, depending on climatic conditions and sunflower genotypes (Toruk et al., 2010). Recommended stages vary between 25% flowers blooming and the final flowering stage (Demiirel et al., 2006). Half the flower area filled with immature seeds can be a signal for harvest. In Nebraska, sunflower intended for silage can be sown as late as July and be used in double-cropping systems after the wheat harvest (Anderson, 2010). After the flowering stage has passed, crude protein level declines and lignin content greatly increases. Once sunflower has been harvested for silage, livestock can enter the field and graze on the leftover stalks.

To harvest seeds, keep an eye out for ripeness. The back of the flower head will turn from green to yellow and the bracts will begin to dry and turn brown; this happens about 30–45 days after bloom and seed moisture is about 35%. Generally, when the head turns brown on the back, seeds are usually ready for harvest. For indoor bouquets, cut the main stem just before its flower bud has a chance to open to encourage side blooms. Cut stems early in the morning. Harvesting flowers during middle of the day may lead to flower wilting. Handle sunflowers gently. The flowers should last at least a week in water at room temperature. Arrange sunflowers in tall containers that provide good support for their heavy heads, and change the water every day to keep them fresh. Cut the head off the plant (about 4 in. below the flower head) and remove the seeds with your fingers or a fork. To protect the seeds from birds, you can cover the flowers with a light fabric such as cheesecloth and a rubber band. Or, you can cut the flower head early and hang the heads upside down until the seeds are dry; hang indoors or in a place that is safe from birds and mice.

16.22 YIELD

Sunflower is a fast-growing crop with high forage yield capacity. In Cuba, fresh matter yields are 450–750 q/ha in 60–70 days in dry conditions, and up to 900 q/ha in Brazil (Penichet Cortiza et al., 2008). The highest seed yield of sunflower obtained was 60 q/ha with optimum soil moisture, the wetted moisture regime created increased in yield by 7.51% over drier moisture regime. The yield reduction was observed in accordance to deferred dates of sowing.

16.23 SEED PRODUCTION

The crop is ready for harvest when top leaves are dry and flowers are shriveled. Heads may be removed with shears or knife. Heads after cutting are sun dried on the threshing floor. Hand threshing can also be done by rubbing seed heads on a metal sheet or beating with sticks. Threshed seed must be dried to 8–10% moisture before storage. The local cultivar gave highest seed production up to 15 q/ha.

16.24 NUTRITIVE VALUE

Sunflower seed meal was superior to both wheat and corn germ meal (three times superior to soybean meal), giving an average gain in weight of 56 g per rat by the end of the same period of time. By the end of 16 weeks, the average gain was 70 g per rat with 5% wheat and corn germ meals and 70% greater, or 119 g per rat, with the sunflower seed meal. The sunflower seed meal as a light gray palatable powder (53% protein) which can be satisfactorily blended with wheat flour or corn meal to make appetizing baked foods. Its high nutritional properties (vitamin content) suggest that sunflower seed may be of much more practical value in human nutrition than hitherto assumed. Weanling rats were divided into different groups with due regard to litter membership, sex, and weight. Each group was fed the same basic ration plus 5–10% of the product to be tested, this product being the sole source of vitamin B complex. The supplements were defatted wheat germ meal, defatted corn germ meal, defatted sunflower seed meal, defatted soybean meal, and Brewer's yeast (control). The growth rate was least rapid with soybean meal.

16.25 UTILIZATION

The composition of the sunflower meal after oil extraction would depend mainly on the seed variety and the extraction method. Protein and crude fiber are the main compounds in sunflower meal. Plant proteins are economic and sustainable alternatives to animal proteins as functional ingredients in food formulations. Oilseeds are the most important source of plant protein preparations. Sunflower seeds are interesting in view of their widespread availability in areas where soy is not produced or is only sparsely produced. The focus of the industry in sunflower cultivation has been put almost exclusively on oil extraction and production. Most of the research activities carried out in recent years have clearly revealed the potential of sunflower proteins as a high value-added component for human nutrition and biofilms production. However, limitation on the production of large quantities to satisfy the growing market demands include the lack of viable bioprocesses that are transferable to industrial scale. Consequently, the main current use of this protein by-product is in animal feed. Process development for production of sunflower-based protein concentrates, isolates, and hydrolysates tailored for specific food applications and the improvement of the functional and rheological properties of the sunflower proteins could expand their market potential.

16.26 SPECIAL FEATURES

The edible sunflower seeds can be eaten raw, cooked, roasted, or dried and ground for use in bread or cakes, as a snack. The seeds and the roasted seed shells have been used as a coffee substitute. Oil can be extracted and used for cooking and soap making. Yellow dyes have been made from the flowers, and black dyes from the seeds. The residue oil cake has been used as cattle and poultry feed, and high quality silage can be made from the whole plant. The buoyant pith of the stalk has been used in the making of life preservers. Wild sunflower is highly branched with small heads and small seeds, in contrast to the single stem and large seed head of domesticated sunflower. Sunflower perfumes are also popular. The modern sunflower kitchens have a large number of sunflower decorated items such as sunflower painted crockery, dinner sets, and living rooms with sunflower wallpaper, sunflower wallpaper art, sunflower wallpaper borders, sunflower rugs, and sunflower pillows. Golden mini sunflowers herald sunny summer days. The carefree sunflower bouquet sends cheery wishes for all those special occasions—birthday, wedding anniversary, new baby, get well, or to simply say thanks.

16.27 COMPATIBILITY

Sunflower has a wide potential sowing window. High yields may be produced from early plantings, yet yields may be reduced by increased pest problems. Soil temperature should be a minimum of 7°C for planting and around 10°C for germination. Lower temperatures will increase the susceptibility of seedlings to diseases such as downy mildew. The main insect problems in sunflower crops include cutworms, sunflower bud moth, sunflower stem weevil, sunflower root weevil, and the sunflower midge. In the future, it is believed that sunflower will be grown in the more arid areas of the world; this trend is predicted to accelerate particularly in the next 10 years. The next trend will be to establish a higher value market for the product and to make sunflower oil more competitive with palm and soybean oil on the world market. Recent developments in the fatty acids in sunflower oils have occurred, oleic acid values have been increased to make the oil more stable. More developments in this area are expected. As with many other major crops, gene transfer is possible in sunflowers. This is the only method of reducing some of the production problems relating to the sunflower crop despite consumer concerns in some countries.

Sunflower is perceived to be a drought-tolerant crop as it roots deeply and extracts water at depths not reached by other crops. Sunflower is comparable to maize in many ways, although it can extract water more efficiently in low-rainfall areas. The seedbed should be prepared so that a moist soil environment is available for germination and growth. The soil surface should be left as rough as possible to reduce the risk of soil erosion; drifting and blowing soil can seriously damage young seedlings. If the soil becomes compacted prior to planting, then reduced aeration and restricted water movement will occur, these conditions will increase the risk of downy mildew occurring. Breakdown of soil structure also reduces nutrient and water uptake and therefore yield. Sunflower has a wide potential sowing window. High yields may be produced from early plantings, yet yields may be reduced by increased pest problems. Nitrogen applications of 50–75 kg/ha are generally sufficient.

FIGURE 16.1 (See color insert.) Sunflower.

KEYWORDS

- **sunflower**
- **origin**
- **morphology**
- **characteristics**
- **cultivation**

REFERENCES

Annon. Tallest Sunflower. *Guinness World Records*, 2016.

Adam, J. A. *Mathematics in Nature*, ISBN 978-0-691-11629-3, 2003.

Anderson, V. Sunflowers are Suitable Late Silage Crops. *Hay and Forage Grower*; Penton Media Inc.: New York, USA, 2010.

Demiirel, M.; Bolat, D.; Celik, S.; Bakici, Y.; Tekeli, A. Evaluation of Fermentation Qualities and Digestibilities of Silages Made from Sorghum and Sunflower Alone and the Mixtures of Sorghum–Sunflower. *J. Biol. Sci.* **2006,** *6* (5), 926–930.

Motloch, J. L. *Introduction to Landscape Design*, ISBN 978-0-471-35291-4, 2000.

Penichet Cortiza, M.; Carballo García, P.; Guerra Gárcés, M.; Alemán Pérez, R. El Cultivo del Girasol como Alternativa Forrajera Viablc Para la Alimcntacion del Ganado Vacuno Lechero. *Observatorio de la Economía Latinoamericana.* 2008; p 95.

Putt, E. D. Early History of Sunflower. In *Sunflower Technology and Production Agronomy Series 35*; Schneiter, A. A., Ed.; American Society of Agronomy: Madison, Wisconsin, 1997; pp 1–19.

Toruk, F.; Gonulol, E.; Kayısoglu, B.; Koc, F. Effects of Compaction and Maturity Stages on Sunflower Silage Quality. *Afr. J. Agric. Res.* **2010,** *5* (1), 055–059.

FIGURE 1.1 Setaria grass.

FIGURE 3.1 Bermuda grass.

FIGURE 9.1 Lathyrus.

FIGURE 11.1 Soybean.

FIGURE 14.1 Green gram.

FIGURE 15.1 Black gram.

FIGURE 16.1 Sunflower.

FIGURE 17.1 Mustard.

FIGURE 20.1 Amaranthus in flowering stage.

FIGURE 21.1 Subabul.

FIGURE 22.1 Gliricidia.

FIGURE 25.1 Various types of pits for *Azolla* cultivation.

CHAPTER 17

BRASSICAS

UTPAL GIRI[1*], SOMA GIRI[2], NAVENDU NAIR[3], ABHIJIT SAHA[1], SONALI BISWAS[4], NILADRI PAUL[5], M. K. NANDA[6], and PROTIT BANDYOPADHYAY[4]

[1]*Department of Agronomy, College of Agriculture, Lembucherra, West Tripura 799210, Tripura, India*

[2]*Horticulture division, Krishi Vigyan Kendra, Ashoknagar, North 24 Parganas 741723, West Bengal, India*

[3]*Department of Entomology, College of Agriculture, Lembucherra, West Tripura 799210, Tripura, India*

[4]*Department of Agronomy, Bidhan Chandra Krishi Viswavidyalaya, Mohanpur, Nadia 741252, West Bengal, India*

[5]*Department of Soil Science and Agricultural Chemistry, College of Agriculture, Lembucherra, West Tripura 799210, Tripura, India*

[6]*Department of Agricultural Meteorology and Physics, Bidhan Chandra Krishi Viswavidyalaya, Mohanpur, Nadia 741252, West Bengal, India*

[*]*Corresponding author. E-mail: utpalagro84@gmail.com*

ABSTRACT

The members of the *Brassica* family such as radish, turnip, swedes, broccoli, brussel sprouts, cauliflower, and cabbage are cultivated as forage crops during winter season in India. However, these crops are generally cultivated as vegetable crops. They are commonly sown in October–November onwards as they require cool temperature, and harvested in spring and early summer when pasture quality is often low or in winter when pasture quantity is limited. Forage brassicas can provide quick and abundant feed, with high digestibility, energy, and protein. Brassicas are high-quality forage if harvested before heading. Aboveground parts normally have 20–25% crude protein and 65–80% total digestable nitrogen. The roots of turnips and kale usually have 10–14% crude protein and 80–85% digestibility.

17.1 BOTANICAL CLASSIFICATION

Kingdom: Plantae
Order: Cruciferae
Tribe: Brassicaceae
Genus: *Brassica*

17.2 BOTANICAL NAME

Leaf turnip: *Brassica rapa*; syn. B. *campestris* var. *Rapa*
Bulb turnip: *Brassica rapa*; syn. B. *campestris* var. *Rapa*
Swedes: *Brassica napus* spp. *napobrassica*
Rape: *Brassica napus* spp. *biennis*
Kale: *Brassica oleracea* L. var. *acephala*
Chinese cabbage: *Brassica campestris* var. *pekinensis*

17.3 COMMON NAME

Leaf turnip or forage hybrids brassica (or turnip cross) (Chinese cabbage × turnip hybrid brassica); bulb turnip; swedes; rape; kale; Chinese cabbage.

17.4 INTRODUCTION

The members of the *Brassica* family such as radish, turnip, swedes, broccoli, brussel sprouts, cauliflower, and cabbage are cultivated as forage crops during winter season in India. However, these crops are generally cultivated as vegetable crops. They are commonly sown in

October–November onwards as they require cool temperature, and harvested in spring and early summer when pasture quality is often low or in winter when pasture quantity is limited. Forage brassicas can provide quick and abundant feed, with high digestibility, energy, and protein. They require relatively low cost for their establishment. Forage brassicas can suppress the weed growth by their rapid initial growth and more vegetative cover. Therefore, it is an excellent cover crop. It appears to be a good tool as a biofumigant cover crop for control of nematodes, soil-borne diseases including the fungal pathogen (*Rhizoctonia*) responsible for damping-off, weeds, and other pests. When biomass is incorporated into the soil, soil microbes break down sulfur compounds in the plant into isothiocyanate, which can act as a fumigant and weed suppressant (Olmstead, 2006). Brassica crops can also reduce the incidence of soil-borne plant diseases by producing naturally occurring chemicals called glucosinolates which breakdown in the soil to produce beneficial compound that inhibit pathogen growth also. The aboveground parts (stems and leaves) of rape and kale and all parts (stems, leaves, and roots) of turnips and swedes are utilized by livestock. Brassicas are high-quality forage if harvested before heading. Aboveground parts normally have 20–25% crude protein and 65–80% total digestable nitrogen. The roots of turnips and kale usually have 10–14% crude protein and 80–85% digestibility.

Brassica crops have the ability to outcompete "bought in feeds" as a cost-effective source of dry matter. It can produce large quantities of feed for a relatively low cost of production. It can be grazed where they are grown, thereby eliminating additional costs associated with hay, silage, and grain. Break the perennial weed cycle using nonselective herbicides leading to weed-free pastures. Break clover pest and disease lifecycles for better clover content in subsequent pastures (Ruiter et al., 2009). Break the wild endophyte cycle for sowing of novel endophyte grass seed. Grow with low water requirement. Have no significant increase in labor requirements.

17.5 TYPES OF FORAGE BRASSICAS

There are mainly six types of forage brassicas cultivated all over the world (Table 17.1). These six types of forage brassicas differ in their characteristics and uses. In general the differences are explained as follows.

TABLE 17.1 Latin Name and Vegetative Description of the Different Types of Forage Brassicas.

Common name	Latin name	Vegetative description
Leaf turnip or forage hybrids brassica (or turnip cross) (Chinese cabbage × turnip hybrid brassica)	*Brassica rapa*; syn. *Brassica campestris* var. *rapa*	Nonbulb producing
		Swollen taproot provides multiple growing points
		Able to regrow after grazing
		Leafy
Bulb turnips	*B. rapa*; syn. *B. campestris* var. *rapa*	Fleshy bulb
		No neck
		Yellow fleshed (hard)
		White fleshed (soft)
Swedes	*Brassica napus* spp. *napobrassica*	Fleshy bulb
		Obvious neck
		White or yellow fleshed
Rape	*B. napus* spp. *biennis*	Numerous leaves
		Fibrous stem
		No bulb or fleshy stem
		Grows to various heights
Kale (Chou Moellier)	*Brassica oleracea* L. var. *acephala*	Large swollen stem with varying leaf percentage
		Stem—woody outer layer, soft fleshy marrow
		Grows to various heights
Chinese cabbage	*B. campestris* var. *pekinensis*	Elongated head (pe-tsai)
		Nonheading type with a rosette of oblong, dark green leaves, thick white petioles, resembling celery, or Swiss chard

17.5.1 FORAGE RAPE

Forage brassicas are also referred to as rape. Most types have a stringent grazing management requirement and must reach maturity before being grazed. Maturity is indicated by a change in leaf color to a purplish or bronze color that occurs 10–14 weeks after sowing. It is a multistemmed crop with fibrous root and stem height, diameter, and palatability vary with the variety. Rape is one of the best crops for fattening lambs and flushing ewes. Yield is maximized with a 180-day growth period for many varieties while most hybrids; on the other hand, produce greatest yields when allowed to grow 60 days before first harvest and 30 days before the second harvest.

17.5.2 LEAFY TURNIPS OR FORAGE BRASSICA HYBRIDS

Grazing leafy turnips can commence earlier than rapes (6–10 weeks), without waiting for the leaf to turn a characteristic bronze or purple color. Leafy turnips can vary in their ability to maintain palatability and leaf quality with age.

17.5.3 KALES

Most of the animals those eat any kind of leafy foods love kale. Kale is the tallest growing of all forage brassicas, and is slower to mature than the hybrid brassicas. It will produce large quantities of leaf and stem which is utilized mostly by cattle. Most kales have a characteristic winter habit and a greater cold tolerance than other brassicas. Kale varieties vary greatly in establishment, stem development, and time requirement to reach maturity. Stemless type reach a height of 25 in., the narrow stem type reach a height of 60 in. with primary stem up to 2 in. in diameter, whereas stemless kale reaches maturity in about 90 days (Masabni, 2011).

17.5.4 TURNIPS

Both turnip leaf and bulbs are utilized by grazing animals. The turnip bulb is a large storage organ that develops in the first year. The dry matter of turnip feed is around 60% for bulb and 40% for leaf, depending on the size/age and variety of turnip. Turnips are best grazed once at maturity (10–16 weeks). Regrowth is possible from a light first grazing if the leaf growing points attached to the bulb are not damaged. Turnips suffer less from insect attack than other forage brassicas; however, they are not as drought tolerant.

17.5.5 SWEDES

Swedes require higher rainfall than turnips. Swedes have a larger bulb than turnips and are slower to mature (20–24 weeks). Swedes are better than turnips at maintaining bulb quality over winter. Swedes usually produce a short stem but can have stems up to 2½-ft long when grown with tall crops which shade the swede. Grazing normally commences during winter and the crop is usually only grazed once. Swede is very high in energy and dry matter content.

17.5.6 CHINESE CABBAGE

It is not a member of cole group but it is regarded as a closely allied crop due to similarity in breeding systems, cultivation practices, and season of growing and other features. Chinese cabbage is a leading market vegetable in China, Japan, and Southeast Asia and grown on more than 500,000 ha. In tropical Africa, Chinese cabbage is common in city markets and is occasionally recorded as vegetable in many countries. It is recorded as a weed in Ethiopia, Kenya, Zimbabwe, and Kenya.

17.6 ORIGIN AND DISTRIBUTION

Brassica genus has 100 species including coles and mustards, native to north temperate parts of the eastern hemisphere.

Leafy turnips—Middle and Eastern Asia
Bulb turnip—Middle and Eastern Asia
Swedes—Mediterranean region

Forage rape—Southwest Europe
Kale—Europe
Chinese cabbage—China and Eastern Asia, heading Chinese cabbage—Eastern Asia

These crops are mainly cultivated in temperate region such as New Zealand, European countries, United States, and Australia in the world and Himachal Pradesh, Jammu and Kashmir, and Uttarakhand in India. Turnip is extensively cultivated in Bihar, Haryana, Himachal Pradesh, Punjab, and Tamil Nadu.

17.7 CLIMATIC REQUIREMENT

Forage brassicas are the crops of temperate climates. Cool and moist climate is most favorable for growing. However, it can also be grown at higher altitude in the tropics or where summers are mild. Crops require about 15–25°C temperature, low humidity for their better growth and development. However, bulbs develop best flavor, texture, and size at a temperature of 10–15°C. The crop growth is optimum at <25°C, while it ceases at <3°C and >35°C (Table 17.2). Rainfall, high humidity, and cloudy weather are not favorable for the crops. The short-day length and cool weather favor proper development of the crops. Excessive cold and frost are harmful to the crop. They require an annual precipitation of 40–100 cm.

TABLE 17.2 Optimum Temperature (°C) Requirement for Growth of Forage Brassicas.

Crop	Optimum temperature (°C)
Leafy turnips	15–22
Bulb turnip	10–15 (bulb development); 15–22
Swedes	18–25
Forage rape	18–25
Kale	15–20, it can withstand −10 to −20
Chinese cabbage	18–25

17.8 SOIL REQUIREMENT

Brassica forages flourish well on wide range of soils especially on those with high moisture-retention capacity, but they cannot survive in waterlogged condition. With good drainage, fertile loam soil having high humus content is best suited for their better growth and development. The extremely light sandy soil or too heavy soils should be avoided. The pH of the soil should be at least 5.6 but ideally between 5.8 and 7.8 (Table 17.3).

TABLE 17.3 Optimum Soil pH Requirement for Forage Brassicas Cultivation.

Crop	pH
Leafy turnips	5.8–7.7
Bulb turnip	5.8–7.7
Swedes	5.8–7.7
Forage rape	6.0–7.5
Kale	6.0–7.5
Chinese cabbage	5.8–7.8

17.9 FIELD PREPARATION

Brassica seeds are very small, so a fine but firm seedbed is desirable. First plowing should be done with a mould board plow, and subsequent two cross plowings with cultivator or harrow followed by planking. Direct drilling is sometimes used but results may be variable. Successful establishment of the crop depends on good seedbed preparation, weed and pest control, soil type, and rainfall following sowing.

17.10 SOWING METHOD

Broadcasting method is very old method, it requires more seeds but still farmers are using this method particularly for forage cultivation. For line sowing, seeds are sown directly either in lines or on ridges. Normally, flat beds are used for sowing but sowing in low lying area or during the rainy season should be done on ridges. Seeds are sown on ridges or rows 15–20 cm apart, while a spacing of 5–7 cm is kept within the plant (Table 17.4). Seeds can be mixed with sand or ash to facilitate uniform sowing. Generally, thinning is done 10–15 days after germination. The optimum sowing depth for brassica seeds is 1.0–1.5 cm. Brassica seeds are very small, therefore to ensure maximum germination, sow no deeper than 2 cm into a firm, moist seedbed. Forage brassicas can be successfully sown with pastures. Brassica sowing rates should be reduced by half to two-thirds when under sowing pastures.

TABLE 17.4 Seed Rate (kg/ha) and Spacing for Line Sowing of Forage Brassicas.

Crop	Sowing method	Seed rate (kg/ha)	Spacing for line sowing
Leafy turnips	Line sowing	3.0–4.0	30–45 cm× 5–15 cm
	Broadcasting	5.0–6.0	
Bulb turnip	Line sowing	0.8–1.0	30 cm × 5–7 cm × 1.5 cm
	Broadcasting	1.0–3.0	
Swedes	Line sowing	0.8–1.0	15–20 cm × 5–7 cm
	Broadcasting	1.0–3.0	
Forage rape	Line sowing	2.5–4.0	15–20 cm × 5–7 cm
	Broadcasting	5–6.5	
Kale	Line sowing	3.0–5.0	15–60 cm × 5–30 cm
	Broadcasting	5.0–5.5	depending on the purpose of the crop
Chinese cabbage	Line sowing	10–12 kg/ha in irrigated land	20 cm × 15 cm
	Broadcasting	20–25 kg/ha in rainfed condition	

17.11 SEED TREATMENTS

Commercial seed treatments are available for use on brassica seeds. These include insecticides for red-legged earth mite and blue oat mite control, fungicides for seedling diseases, and molybdenum (a trace element often deficient in acid soils). Broadcast and direct drilled crops may be more susceptible to seed theft by birds and may require a bird repellent. Seeds should be treated for control of seed-borne diseases with fungicides such as Bavistin @ 2 g/kg of seeds or mancozeb 75% WP @ 2 g/kg of seed or *Trichoderma viride* @ 5 g/kg of seed.

17.12 PLANTING TIME

In the temperate region of India such as Jammu and Kashmir, Himachal Pradesh, hilly region of Uttarakhand, and Uttar Pradesh, the forage brassicas are sown in the month of March–April and the grazing can be continued to July–August. For better forage production, the crop should be sown as early as possible to avail the extended period of low temperature which is required for better vegetative growth. With the increase in temperature, the flowering will be initiated, particularly in the plains of India. However, the cropping sequence should be selected as per the requirement of the farmers (Table 17.5).

TABLE 17.5 Sowing Time for Forage Brassicas Cultivated in Plains of India.

Crop	Sowing time
Leafy turnips	October–December
Bulb turnip	Asiatic turnips are sown from July to September
	European type is sown in October–December
Swedes	November–December
Forage rape	October–December and February–March
Kale	November–December
Chinese cabbage	November–December

17.13 VARIETIES (Table 17.6)

TABLE 17.6 List of the Varieties of Forage Brassicas.

Crop	Varieties
Leafy turnips	Hunter, Pasja, Tyfon
Bulb turnip	Early Milan Red Top, Pusa Swarnima, Pusa Sweti Golden Ball Punjab, Safed 4, Purple Top White Globe, Pusa Chandrima, Snow Ball, and Pusa Kanchan
Swedes	Aparima Gold, Dominion, Doon Major, Highlander, Invitation, Major Plus, Keystone Winton Virtue, Champion Purple Top, Caldon, and Sensation
Forage rape	Greenland, Giant Titan Wairoa Winfred, Hobson, Hungry Gap, and Interval (Hyb.)
Kale	Dwarf Blue Curled Scotch, Dwarf Blue Curled Vates, Green Curled, Nero di Toscano, Rebor, Red Russian, Dwarf Green Curled Scotch, Dwarf Moss Curled, Marrow Stem, and Gruner
Chinese cabbage	Tropical Delight, Tropical Prince, Tropical Queen, Nozomi, and Optico

17.14 FERTILIZER REQUIREMENT

The quantity of manures and fertilizers to be applied depends upon climate, fertility status, pH level, and texture of the soil. The proper time of application is equally important to facilitate optimum intake of plant nutrients for good harvest. Different doses of NPK have been recommended for various agroclimatic zones of India. However, a basal dose of 20–25 t/ha of farmyard manure should be applied at the time of land preparation.

Brassicas are grown on soils with a wide range of soil fertility. There are no standard recommendations for fertilizer management because each crop has a different yield potential and therefore different nutrient requirement. Phosphorus is the main element required by brassica crops. Phosphorus rates of at least 40–60 kg/ha are suggested depending on soil phosphorous levels, and soil type. Brassica crops are often sown with compound or starter fertilizers containing varying amounts of nitrogen. However, care is needed, as too much N placed close to the seed at sowing can inhibit germination of the crop particularly in dry conditions. Approximately 100 kg/ha of N is commonly used and this can increase both yield and crude protein content. Care must be taken when grazing the crop after N application due to increased risk of nitrate poisoning. Applications of N are recommended at 4–6 weeks and again at 8–12 weeks after emergence depending on the duration of crop as well as variety.

In highly acidic soils, molybdenum deficiency is common. A fertilizer or seed treatment containing molybdenum to supply 50–100 g/ha can be used to overcome deficiencies. Alternatively, liming to increase soil pH may also overcome molybdenum deficiency. Boron deficiency has been known to occur relatively rare particularly in recently limed soils or soils with a high pH. In situations where boron deficiency is suspected or has occurred, a boron fertilizer should be used to supply approximately 2 kg/ha (Table 17.7).

TABLE 17.7 Fertilizer Dose (kg/ha) for Forage Brassicas.

Crop	Fertilizer dose (kg/ha) $N:P_2O_5:K_2O$
Leafy turnips	69–98:44:44
Bulb turnip	70–100:50:50
Swedes	120–160:90–150:250–370
	Boron (B/ha) 1.5 kg

TABLE 17.7 *(Continued)*

Crop	Fertilizer dose (kg/ha) N:P_2O_5:K_2O
Forage rape	150–225:60:60
	Top dressing—75 kg/ha N
Kale	220–240:100–160:100–160
	½ applied preplant with the remaining at thinning (12–15 DAS)
Chinese cabbage	200–220:100–150: 00–250
	MgO—50–100 kg/ha

DAS, days after sowing.

17.15 WATER MANAGEMENT

The crop needs frequent irrigations for realizing the yield potential of the crop. With irrigation, yield responses to applied water are variable, depending on season and location. Irrigation requirements will depend on the prevailing conditions. During the dry months of April–June, crop requires irrigation at an interval of 7–10 days, if required and 15–20 days interval in the month of October–March. The water requirement of these crops varies from 380 to 500 mm.

17.16 WEED MANAGEMENT

Brassicas are especially susceptible to weed competition during the early crop development phases. Two or three weeding regimes at 20 days interval are required to control the weeds. A single spray with a high rate of roundup 2−4 days before plowing is usually sufficient for a good control of weed. Preemergence application of Treflan at 2 L/ha on a light soil or 3 L/ha on a heavy soil may be recommended. Alachlor (4 L/ha) applied preemergence can be used to control weeds. The preemergence application of herbicide, Tok E-25 (Nitrofen) @ 2 kg/ha effectively controls the weeds. Apply Dicamba at 700−850 mL/ha for control of broad leaf weeds. Note, this should not be applied to bulbing crops such as turnips and swedes, and will check growth of rape and kale if applied at higher rates.

17.17 INSECT-PEST AND DISEASE MANAGEMENT

Forage brassicas are attacked by several pests and diseases among which few are very destructive, but the intensity varies from year to year depending on the prevailing weather condition and the crop history. Some of the major pests and diseases, their causal organisms, damage symptoms, and control measures are given below:

17.17.1 INSECT-PESTS

17.17.1.1 DIAMONDBACK MOTH: Plutella xylostella (PLUTELLIDAE: LEPIDOPTERA)

Full grown larvae are tapering toward both ends and measure about 1–1.5 cm in length. Larvae are greenish in color with short thin hairs on the body. Caterpillars feed on green tissues by scrapping from under surface of leaves leaving the upper epidermis intact. Severely infested plants show withered appearance due to presence of numerous holes on leaves. Foliar spray with Lufenuron 5.4% EC @ 240 mL in 200 L of water/acre or spinosad 2.5% SC @ 240–280 mL in 200 L of water/acre is effective.

17.17.1.2 CRUCIFEROUS LEAF WEBBER: Crocidolomia binotalis (PYRALIDAE: LEPIDOPTERA)

Larvae are pale violet in body color with red head, brown longitudinal stripes, and rows of tubercles on its body. Caterpillars web together the foliage and feed from within. Removal and destruction of the webbed leaves along with larvae inside is effective. Chemical control is same as in case of diamondback moth.

17.17.1.3 APHID: Lipaphis erysimi (APHIDIDAE: HEMIPTERA)

They are soft bodied, pear shaped, small insects, and pale-greenish in color, winged or wingless. Both the nymphs and adults remain in colonies and suck cell sap from tender plant parts. Infested parts get deformed and the crop yield is drastically reduced. Installation of yellow sticky traps and foliar spray with dimethoate 30% EC @ 300 mL in 200 L of water/acre or acetamiprid 20 % SP @ 40 g in 200 L of water/acre are effective in managing the pest.

17.17.1.4 TOBACCO CATERPILLAR: Spodoptera litura (NOCTUIDAE: LEPIDOPTERA)

Stout-bodied full grown caterpillars measure about 35–40 mm in length and velvety black in color with some light color markings. The young larvae are gregarious in nature and scrape the green tissues of leaf lamina giving a papery appearance, whereas the grown-up larvae voraciously feed on leaves causing extensive damage. Collection and destruction of egg masses and gregarious larvae at initial stages is very effective. Spraying with indoxacarb 15.8% EC @ 200 mL in 200 L of water/acre or emamectin benzoate 5% SG @ 100 g in 200 L of water/acre can also control the pest effectively.

17.17.1.5 CABBAGE BUTTERFLY: Pieris brassicae (PIERIDAE: LEPIDOPTERA)

Full-grown larvae are greenish in color with black dots and measure about 40–50 mm in length. The young larvae feed on leaves gregariously, whereas the grown-up larvae get dispersed and feed on leaves from margin inwards leaving intact the main veins resulting in skeletonized leaves. Management is same as in case of tobacco caterpillar.

17.17.2 DISEASES

17.17.2.1 CLUB ROOT

It is caused by *Plasmodiophora brassicae.* Infection of the disease results in club formation in roots which interferes with root functioning restricting water and nutrient uptake. Application of lime @ 2.5 t/ha to raise the soil pH slightly above neutral and seed treatment with Thiram @ 2 g/kg seed are effective.

17.17.2.2 BLACK LEG

It is caused by *Phoma lingam.* Whole root system of the infected plants get decayed leading to falling down of the plants. Seed treatment with Thiram @ 2 g/kg of seed can be done to get rid from the disease.

17.17.2.3 DAMPING OFF

It is caused by *Pythium aphanidermatum* and *Fusarium* spp. Infection by the pathogens results in rotting of collar portion at the ground level leading to death of young seedlings. Soil drenching with Captan 75% WP @ 1000 g in 400 L of water/acre or seed treatment with captan 75% WP @ 2–3 g/kg seed can be done to prevent the disease.

17.17.2.4 ALTERNARIA LEAF SPOT

It is caused by *Alternaria* spp. Circular spots with concentric rings are formed on the leaves which then coalesce leading to blighting of leaves. Spraying with zineb 75% WP @ 600–800 g in 300–400 L of water/acre or mancozeb 75% WP @ 600–800 g in 300 L of water/acre is recommended to control the disease.

17.17.2.5 WHITE RUST

It is caused by *Albugo candida.* Patches of white powdery substance are observed on the under surface of the leaves and floral parts become distorted. Spraying with mancozeb 75% WP @ 600–800 g in 300 L of water/acre is suggested to control the disease.

17.18 HARVESTING AND YIELD

Crops are harvested either by cutting them by sickle or by uprooting them along with tubers, depending on the crop and growing condition. Rapid cooling after harvest is must for the prevention of wilting and maintenance of quality of kale crop (Table 17.8).

TABLE 17.8 Duration and Forage Yield (t/ha) of Forage Brassicas.

Crop	Duration	Forage yield (t/ha)
Leafy turnips	45–60 days (6–8 weeks)	2–8
Bulb turnip	65–70 days (early)	2–12
	80–100 days (medium)	19–25
	90–120 days (late)	20–40
Swedes	125–140 days (18–20 weeks)	5–20
Forage rape	80–100 days (12–14 weeks)	3–10
Kale	60–90 (early)	5–20
	110–140 days (long)	10–25
Chinese cabbage	Eight leaf stage after 60–95 days. Harvesting commences usually in January and continues up to February and until the peduncle has elongated and the first flower has opened	25–50

17.19 GRAZING

Brassicas can be harvested for green chop or silage but are most frequently grazed. Grazing management is important to optimize the true potential of these crops. Strip grazing small areas of brassica at a time provides the most efficient utilization. Grazing large areas increases trampling and waste of the available forage. Rape is more easily managed for multiple grazing than are the other *Brassica* species. Grazing can begin when the forage is about 12-in. tall (70–90 days after planting). The pasture should be grazed for a short time period and the livestock removed to allow the brassica to regrow. Rape may be grazed to 10-in. stubble and one to four grazing periods may occur, depending on planting date and growing conditions. Approximately 6–10 in. of stubble should remain after grazing rape to promote rapid regrowth. Regrowth may be grazed in as few as 4 weeks after the first grazing (Smyth, 2015). Graze rape close to ground level during the final grazing. When turnips are to be grazed twice only the tops should be grazed during the first grazing. Turnip regrowth is initiated at the top of the root, so this part of the plant should not be removed until the second and final grazing when the whole plant can be consumed. Like rape, regrowth of turnips can be sufficient to graze again within 4 weeks of the first grazing.

Brassica crops can cause animal health disorders if not grazed properly. The main disorders are hypothyroidism, and polioencephalomalacia. The disorders can be avoided by the following two management practices:

1. Introduce grazing animals to brassica pastures slowly (over 3–4 days). Avoid abrupt changes from dry summer pastures to lush brassica pastures. Do not turn hungry animals that are not adapted to brassicas into a brassica pasture.
2. Brassica crops should not constitute more than 75% of the animal's diet. Supplement with dry hay if continually grazing brassicas or allow grazing animals to access grass pastures while grazing brassicas.

The forage quality of brassica is so high that it should be considered similar to concentrate feeds and precautions should be taken accordingly. Livestock should not be hungry when to be put on pasture the first time, so they do not gorge themselves. A lower quality hay should be made available to provide some fiber in the animals' diet.

Brassicas can provide grazing at any time during the summer or winter depending on seeding date. A promising use may be for late winter grazing. These crops maintain quality, if not heading, well into freezing temperatures and may be grazed into November (Table 17.9).

TABLE 17.9 Planting and Grazing Sequence for Forage Brassicas.

Crop	Plant part consumed	Seeding to harvest (days)	Regrowth after harvest
Kale	Herbage	150–180	No
Rape	Herbage	80–90	Yes
Swede	Herbage and root	150–180	No
Turnip	Herbage and root	80–90	Yes
Chinese cabbage	Herbage	60–95	Yes

17.20 ANIMAL HEALTH ISSUES

Livestock health problems from grazing brassicas are relatively rare and can largely be avoided by good agronomic and grazing management

(Ayres, 2002). Some livestock health problems that are known to occur include the following.

17.20.1 PHOTOSENSITIZATION

Grazing crops too early, prior to maturity, can cause animals to suffer from photosensitization. Young animals (especially lambs) are prone to photosensitization, while animals with dark pigmented skins and wool covering are much more tolerant. The most common sign of photosensitization occurs on unprotected body parts such as the face and ears. Swelling occurs followed by blistering and scabbing of the ears and face. Rapes and kales are most commonly associated with the disorder, while turnips, swedes, and hybrid brassicas are less likely to cause photosensitization.

17.20.2 NITRATE POISONING

Nitrates accumulate in plant leaves, and in very high concentrations may cause livestock death. This problem is largely caused by high soil nitrate levels following prolonged dry conditions being quickly taken up with rapid growth following rainfall or irrigation.

17.20.3 GOITER (ENLARGED THYROID)

This is sometimes a problem in young lambs, where pregnant ewes have been grazing leafy brassica crops.

17.20.4 KALE ANEMIA

This disorder (sometimes referred to as red water) can occur with all brassica crops, but is more common with kale crops. Anemia is caused by excess levels of the amino acid compound S-methyl cysteine sulphoxide (SMCO) in the plant. SMCO causes a decrease in hemoglobin concentration and a depression of appetite.

Digestive disturbances, respiratory problems, blindness, pulpy kidney, choking, etc. are some other problems which may occur in livestock during their grazing.

Brassicas vary greatly in maturity, bulb shape, leaf to stem ratio, and winter hardiness. The choice of brassica and variety will depend on when the crop will be grazed, livestock class or type, and if multiple grazing is required.

FIGURE 17.1 **(See color insert.)** Mustard.

KEYWORDS

- *Brassica*
- **agronomic package and practices**
- **fodder yield**
- **nutritive value**
- **animal health issues**

REFERENCES

Ayres, L. Forage Brassicas: Quality Crops for Livestock Production. District Agronomist, Orange Bruce Clements, District Agronomist, Bathurst, Agfact First Ed. [online] 2002, P 2.1.13 (accessed Dec 15, 2016).

Masabni, J. Collards/Kale; Commercial and Specialty Crop Guides Department of Horticulture Texas AgriLifeExtensionServicehttp://aggie.horticulture.tamu.edu/vegetable/files/2011/10/collardskale.pdf (accessed Dec 26, 2016).

Olmstead, M. A. Cover Crops as a Floor Management Strategy for Pacific Northwest Vineyards. EB2010. Washington State Univ. Extension, Prosser, [online] 2006, http://cru.cahe.wsu.edu/CEPublications/eb2010/eb20 20.pdf (accessed Dec 15, 2016).

Production Guidelines of Chinese Cabbage. Department of Agriculture, Forestry and Fisheries, 2013, http://www.nda.agric.za/docs/Brochures/chinese.pdf (accessed Dec 26, 2016).

Ruiter, J. de.; Wilson, D.; Maley, S.; Fletcher, A.; Fraser, T.; Scott, W.; Berryman, S.; Dumbleton A. and Nichol, W. Management Practices for Forage Brassicas. Forage Brassica Development Group New Zealand Institute for Plant & Food Research Limited, Lincoln University, Ag. Research and PGG Wrightson Seeds, [online] June, 2009, http://bal.preprod.intergen.net.nz/Documents/Farm/Management%20practices%20for%20forage%20brassicas.pdf (accessed Dec 26, 2016).

Smyth, S. Root & Forage Crops. Mole Valley Farmers Arable Department, [online] 2015, www.molevalleyfarmers.com (accessed Dec 26, 2016).

CHAPTER 18

TURNIP (SALGAM)

PARVEEN ZAMAN[1*] and MD. HEDAYETULLAH[2]

[1]*Assistant Director of Agriculture, Pulse and Oilseed Research Sub-station, Department of Agriculture, Government of West Bengal, Beldanga 742133, Murshidabad, West Bengal, India*

[2]*AICRP on Chickpea, Directorate of Research, Bidhan Chandra Krishi Viswavidyalaya, Kalyani 741235, Nadia, West Bengal, India*

[*]*Corresponding author. E-mail: parveenzaman1989@gmail.com*

ABSTRACT

Turnip is popularly known as salgam. This root crop is palatable, succulent, easily digestible, and a popular feed of all livestock. The turnip or white turnip is a root vegetable commonly grown in temperate climates worldwide for its white, bulbous taproot system. Small, tender varieties are grown for human consumption, while larger varieties are grown as a feed and fodder for livestock. The average yield of turnip fodder is 400–600 q/ha under good management conditions. The feeding of freshly harvested roots can cause scouring in the livestock. When boiled turnips are fed to the animals, they may suffer from nitrite toxicity.

18.1 BOTANICAL CLASSIFICATION

Kingdom: Plantae
Orders: Brassicales
Family: Brassicaceae
Genus: *Brassica*
Species: *rapa* L.

Variety: *rapa*
Binomial name: *Brassica rapa* var *rapa* L.

18.2 BOTANICAL NAME

Brassica rapa var *rapa* L.

18.3 COMMON NAME

Turnip, salgam.

18.4 INTRODUCTION

Turnip or white turnip belongs to mustard family and is popularly known as salgam. This root crop is palatable, succulent, easily digestible, and popular feed of all livestock. The turnip or white turnip is a root vegetable commonly grown in temperate climates worldwide for its white, bulbous taproot system. Small, tender varieties are grown for human consumption, while larger varieties are grown as a feed and fodder for livestock. Turnip is profusely used as a fodder for livestock worldwide. Turnip leaves are sometimes eaten as vegetables, like mustard. Turnip greens are a common side dish in many countries. Smaller leaves are preferred as vegetables but the bitter taste of larger leaves can be reduced by pouring off the water from the initial boiling and replacing it with fresh water. The feeding of freshly harvested roots can cause scouring in the livestock. When boiled turnips are fed to the animals, they may suffer from nitrite toxicity.

18.5 ORIGIN AND DISTRIBUTION

Turnip is a native of Eurasia. It is now grown in Europe, the United States, New Zealand, Canada, Australia, the Middle East, and Far East countries. It is grown as a fodder in the states of North and West India but as a vegetable, it is grown throughout the country. Some evidence shows that the turnip was domesticated before the 15th century B.C.; it was grown in India at this time for its oil-bearing seeds. The turnip was a well-established crop in Hellenistic and Roman times, which leads to the assumption that it was brought into cultivation earlier. The wild forms of the hot turnip and its relatives, the mustards and radishes, are found in West Asia and Europe.

18.6 PLANT CHARACTERISTICS

The turnip stem is short and elongates slowly to form the floral shoot. The leaves are hairy and the taproot is swollen and napiform type. The seeds are globose and borne in two-celled pods. The most common type of turnip is mostly white skinned apart from the upper 1–6 cm, which protrude above the ground and are purple or red or greenish where the solar radiation comes in contact. This aboveground part develops from stem tissue but is fused with the root. The interior flesh is entirely white. The root is roughly globular, from 5 to 20 cm in diameter, and lacks side roots. Underneath, the taproot is thin and 10 cm or more in length. The leaves grow directly from the above the ground shoulder of the root, with little or no visible crown or neck.

18.7 CLIMATIC REQUIREMENT

Turnip grows best in cool and moist climate. The temperature ranges between 18 and 28°C during the growing period. It is cultivated throughout the mild winter in the plains and Central India. It is also a rainfed crop during the monsoon. Hot and dry summer and freezing conditions should be avoided. Highly fertile and well drained loam is the best soil for its growth. Sandy soils require less manuring and irrigation. The crop tolerates salinity even up to 8 mmhos/cm. Clay soils are not preferred.

18.8 SOIL AND ITS PREPARATION

The root crop requires deep, loose, and friable seedbed for easy development of the taproot. One deep plowing with a country plow, followed by three to four operations with a country plow or cultivator and disc harrows, is necessary to prepare the desired tilth. Soil moisture conservation is important for successful establishment of turnips.

18.9 VARIETIES

Some important varieties of turnip are Appin, Barabas, Barkant, Marco, Purple Top, Polybra, Rondo, Samson, Vollenda, etc. for high-yielding fodder crops (Jacobs et al., 2001).

18.10 SOWING TIME

In the plains, it may be from September to November and in north Himalayan region it may be from April to June, after snow melts away. The seeds are either sown as: broadcast or in lines. The average sowing date recorded on a survey of 142 commercial farms was mid-October to 17th December (Jacobs et al., 2001). A study by Jacobs et al. (2002) in Southwest Victoria showed that, sowing early (mid-October) rather than later (mid-November) ensured higher soil moisture at sowing time.

18.11 SEED TREATMENT AND INOCULATION

To treat the seed with fungicide apply Thiram @ 3 g/kg of seed. Seeds should be inoculated with bacteria before sowing for fixation of atmospheric nitrogen. Apply 20 g each *Azotobacter* and phosphorus solubilizing bacteria for 1 kg of seed before sowing for nitrogen fixation.

18.12 SEED RATE AND SOWING METHOD

The seed rate for normal sowing is varying from 3 to 5 kg/ha. When grown in association with oats for fodder, this rate may be reduced to half. Sowing rates from 1.6 to 2.2 kg/ha has been used in Southern Victoria, reaching seedling densities from 28 to 45 plants/m^2 (Jacobs et al., 2001). Lower rates (0.5 kg/ha) encourage bulb development and high rates (up to 2 kg/ha) allow early yield and a high leaf:bulb ratio.

18.13 SEED BED

Turnips should be sown shallow (5–10 mm) in a loose, friable, and moist seedbed. Rolling after sowing to improve the seed–soil contact is a normal practice with this species. They can also be suitable for direct drilling in friable soils, although a successful establishment is more difficult with this technique.

18.14 SPACING

In line sowing, row-to-row spacing is maintained about 30 cm and between plants 10 and 15 cm; but depending upon the soil, climate, and irrigation facilities it may be increased. In dry areas, the spacing may be increased up to 50 cm when it is grown in intercropping systems with oats and barley.

18.15 CROP MIXTURE

Turnip crop can be grown in association with oats, wheat, barley, gram, etc.

Some of the important rotations of the turnip are given bellow:

- Sorghum–turnip
- Maize–turnip
- Pearl millet–turnip

18.16 NUTRIENT MANAGEMENT

Farmyard manure or compost may be used at the rate of 10–15 t/ha at the time of land preparation as a basal. Basal application of N at the rate of 25 kg/ha in the form of ammonium sulfate to provide sulfur to the crops. Sulfur is essential for its growth and development. Turnips are normally sown with 20–25 kg P_2O_5/ha as: triple superphosphate, monoammonium phosphate, and diammonium phosphate. It is important to ensure that molybdenum (Mo) and boron (Bo) are not deficient (Mo can be provided with seed coating). Turnips can adapt to a broad range of pH; they perform better on soils with a pH more than 5.5.

18.17 WATER MANAGEMENT

The seed requires high moisture or one irrigation immediately after sowing for quick germination. After germination, the taproot should be allowed to go into deep. Turnip requires three to four irrigations. If rain fails then supplementary irrigation may be required for better growth.

18.18 WEED MANAGEMENT

Turnip fodder crop requires weeding very early. Crop–weed competition is more common in early stage of crop growth. This is a smothering crop; it suppresses the weed and covers the land.

18.19 INSECT-PEST AND DISEASE MANAGEMENT

In a survey by Jacobs et al. (2001), insect damage had the highest impact on total yields compared to other factors such as total water received, soil temperature, soil moisture, and seedling density. Several pests affect turnips: red legged earth mite, slugs, aphids, cabbage moth, cabbage white butterfly, diamondback moth, cutworms, lucerne flea, wingless grasshoppers, and leaf miners. The diamondback moth is the most common and damaging pest of *Brassica* crops and has had a severe impact in forage turnip crops of dairy farms in Southern Victoria.

18.20 HARVESTING

The forage harvesting is done by hand lifting at the age of 8–10 weeks. The fodder can be harvested from the month of November up to April.

18.21 YIELD

Turnip is the most quickly growing fodder crop. The average yield of turnip fodder is 400–600 q/ha under good management conditions. The fodder yield may be as high as 700–800 q/ha under best management package and practices.

18.22 SEED PRODUCTION

Turnip requires cooler climate for its seed production. For seed production, entire leaves are harvested with a height of 5–6 cm in the months of January–February. Then, the bottom one-third portions of the root are chopped off. The top portions of the roots with trimmed leaves are then planted in a well prepared field. The spacing is maintained in 50–60 cm row-to-row and 25–30 cm plant to plant. Then the crop is required frequent light irrigation every 3–4 weeks. The flowers come in the month of April and are harvested in May. The average seed yield may vary from 6 to 8 q/ha (Chatterjee and Das, 1989).

18.23 NUTRITIVE VALUE

On dry matter basis, the roots contain 14–18% crude protein and 40% sugar. The turnips are quite rich in vitamins and minerals. The turnip root is rich in vitamin C. The green leaves of the turnip top are a good source of vitamin A, vitamin C, vitamin K, and calcium. Turnip greens are also high in lutein (8.5 mg/100 g).

18.24 TOXICITIES

The feeding of freshly harvested roots can cause scouring in the livestock. When boiled turnips are fed to the animals, they may suffer from nitrite toxicity.

18.25 UTILIZATION

Turnip is profusely used as a fodder for livestock worldwide. Turnip leaves are sometimes eaten as vegetables, like mustard. Turnip greens are a common side dish in many countries. Smaller leaves are preferred as vegetables but the bitter taste of larger leaves can be reduced by pouring off the water from the initial boiling and replacing it with fresh water. Most baby turnips can be eaten whole, including their leaves. Their flavor is mild, so they can be eaten raw in salads.

KEYWORDS

- **turnip**
- **agronomic management**
- **fodder yield**
- **nutritive value**
- **utilization**

REFERENCES

Chatterjee, B. N.; Das, P. K. *Forage Crop Production Principles and Practices*; Oxford & IBH Publishing Co. Pvt. Ltd.: Kolkata, 1989.

Jacobs et al. A Survey on the Effect of Establishment Techniques, Crop Management, Moisture Availability and Soil Type on Turnip Dry Matter Yields and Nutritive Characteristics in Western Victoria. *Aust. J. Exp. Agric.* **2001,** *41*, 743–751.

Jacobs et al. Effect of Seedbed Techniques, Variety, Soil Type and Sowing Time, on *Brassica* Dry Matter Yields, Water Use Efficiency and Crop Nutritive Characteristics in Western Victoria. *Aust. J. Exp. Agric.* **2002,** *42*, 945–952.

CHAPTER 19

GAJAR (CARROT)

MOHAMMED ABDEL FATTAH[1] and MD. HEDAYETULLAH[2*]

[1]*Department of Horticulture, Faculty of Agriculture, Cairo University, Giza Egypt*

[2]*Department of Agronomy, Bidhan Chandra Krishi Viswavidyalaya, Mohanpur, Kalyani 741235, Nadia, West Bengal, India*

[*]*Corresponding author. E-mail: bogey1992@gmail.com*

ABSTRACT

Carrot is popularly known as *gajar* in India. This root crop is palatable, succulent, easily digestible, and popular feed of all livestock. The feeding value is superior to other root forage crops. The carrot takes about 3 months to be ready for harvest. Harvesting can start after last irrigation by pulling by hand, a spade, or a country plow. Carrots are harvested when they have reached a diameter of 20 mm and more, still young and tender. The carrot is a root vegetable commonly grown in temperate climates worldwide for its white, bulbous taproot system. Small, tender varieties are grown for human consumption, while larger varieties are grown as a feed and fodder for livestock. Carrot leaves are good poultry feed. Carrot is the most quickly growing fodder crop. The average yield of carrot fodder is 500–600 q/ha under good management conditions. The average seed yield may vary from 6 to 8 q/ha.

19.1 BOTANICAL CLASSIFICATION

Kingdom: Plantae
Orders: Apiales
Family: Apiaceae or Umbelliferae

Genus: *Daucus*
Species: *Carota*
Variety: *sativa* DC
Binomial name: *Daucus carota* var *sativa*

19.2 BOTANICAL NAME

Daucus carota var *sativa*

19.3 COMMON NAME

Carrot.

19.4 INTRODUCTION

Carrot belongs to Apiaceae family and is popularly known as *gajar* in India. This root crop is palatable, succulent, easily digestible, and popular feed of all livestock. The feeding value is superior to other root forage crops. Carrot is a root vegetable commonly grown in temperate climates worldwide for its white, bulbous taproot system. Small, tender varieties are grown for human consumption, while larger varieties are grown as a feed and fodder for livestock. Carrot takes about 3 months to be ready for harvest. Harvesting can start after last irrigation by pulling by hand, a spade, or a country plow. Carrots are harvested when they have reached a diameter of 20 mm and more, still young and tender. Carrot leaves are good poultry feed. Carrot is the most quickly growing fodder crop. The average yield of carrot fodder is 500–600 q/ha under good management conditions. The tuber dry matter contains about 12–14% of which 1% crude protein, 9% carbohydrate, and 1% fiber. The carrot leaves are quite rich in calcium, phosphorus, and minerals.

19.5 ORIGIN AND DISTRIBUTION

Carrot is a native of India and Eurasia. It is now grown in Europe, the United States, New Zealand, Canada, Australia, the Middle East, and Far

East countries. It is grown as a fodder in the states of North and West India but as a vegetable, it is grown in throughout the country.

19.6 PLANT CHARACTERISTICS

The carrot is an erect, annual, or biennial plant, 30–100 cm in height. The root is the edible part and it is basically a swollen base of the taproot that also includes the hypocotyls. It is conical and its length varies from 5 to 25 cm. The color of the roots varies from white, yellow, orange–yellow, light purple, deep red to deep violet. The stem consists of a small plate-like "crown." Leaves are produced in the first season. They have long petioles and are pinnately compound flower. The inflorescence is a terminal compound umbel, subtended by pinnatifid bracts. The flowers of the umbel are white except for the central ones which are either red or purple. Flowers are produced in the second year. Seeds, like flowers, are produced in the second year.

19.7 CLIMATIC REQUIREMENT

The carrot is a cool-weather crop and it also does well in warm climates. The optimum temperature for growth is between 15°C and 20°C. Temperatures below 10°C cause longer, more slender, and paler roots. Shorter, thicker roots are produced at higher temperatures. Extended periods of hot weather can cause strong flavor and coarse roots. Development is also slower in winter than in spring and summer. Carrots require a steady supply of moisture and it must be maintained at above 50% of available moisture throughout growth. Generally, carrots require approximately 25 mm of water per week but under warm, dry conditions 50 mm will be required and it also depends upon the soil type.

19.8 SOIL AND ITS PREPARATION

Carrots require a well-drained fine-textured or sandy soil to prevent forking. Heavy soils and soil compaction can cause deformation and stunting in roots (Sorensen, 2000). Carrots tolerate a wide range of pH values ranging 4.2–8.7 but prefer a pH from 5.5 to 7.0 (OSU, 2004). Apply lime only if the soil pH is lower than 5.2 (OSU, 2004). Carrot prefers deep, loose, and sandy to loamy soils with a pH of 6.0–6.5. The soil should be well plowed and as level as possible in order to obtain a good stand. It must have a good crumbly structure and kept moist enough to allow seed germination. Therefore, the soil must be deep plowed to loosen the soil to a depth of at least 30 cm.

19.9 VARIETIES

Fodder varieties are generally large in size, white fleshed, and yield more than the vegetable varieties. The popular varieties for fodder purposes are Sirsa, Belgian, Improved Long Orange, Mastodon, and Giant yellow oxheart. Many varieties, both indigenous and exotic, differing in temperature requirement, length, size, shape, and color of roots and duration of crop are grown in India. Tropical or Asiatic or annual types do not require low temperature for flowering and they produce seeds in plains of North India, for example, Pusa Kesar and Pusa Meghali (Chatterjee and Das, 1989).

19.10 SOWING TIME

Sowing of carrot starts from September to November and it may be from April to June after snow melts away in temperate regions.

19.11 SEED RATE AND SOWING METHOD

The seeds are directly sown in the field on ridges or raised beds. Row planting is preferred to broadcast sowing. The seeding depth should be 10–25 mm or 40 mm in loose, light sand. Planting depth should be shallow on heavier soils and in colder months. Slightly deeper planting is recommended in summer when the soil dries out quickly. The chance of a successful establishment of the crop will be increased.

19.12 SPACING

The rows are generally spaced from 20 to 40 cm apart. A planting density of 180–190/m^2 gives good results in double rows whereas a density of 130/m^2 is ideal for single-row fodder crops.

19.13 CROP MIXTURE

Carrot can be grown in association with oats, wheat, barley, gram, etc. Some of the important rotations of the carrot are given bellow:

- Sorghum–carrot
- Maize–carrot
- Pearl millet–carrot
- Paddy–carrot
- Jute–carrot

19.14 NUTRIENT MANAGEMENT

Compost or organic manure should not be applied since they cause unattractive, hairy roots, with a coarser texture. Fertilizer recommendations should be based on soil analyses. A fertilizer dose of 40–50 kg N, 40–50 kg P_2O_5, and 80–100 kg K_2O is recommended for the crop. As a general guide, 100 kg/ha fertilizer mixture of 2:3:4 should be worked into the soil before planting. Limestone ammonium nitrate at a rate of 100–200 kg/ha should be applied as a top dressing 8 weeks after planting. Carrots have low nitrogen requirements, and good yields can be obtained with 80 kg/ha of nitrogen applications. Phosphorus at the rate of 40 kg/ha are sufficient to produce a good crop. The crop has a high potassium requirement and a half is applied as a side dressing at 4–8 weeks after planting. The balance is applied as a late dressing.

19.15 WATER MANAGEMENT

The soil should never be allowed to dry out. Too much moisture causes short carrots with light color and a larger diameter. The field should be irrigated lightly immediately after sowing. Irrigation water should be applied once or twice a day using a sprinkler system. Watering should gradually be reduced to prevent longitudinal splitting of the roots when the crop approaches maturity. Water stress during root development also causes cracking of the roots, which also become hard. First irrigation should be applied immediately after sowing followed by another 4–6 days after. Soil should be kept moist by frequent light irrigation for proper growth of roots. Excessive irrigation, that too towards the last stage, should be avoided as it may result in excessive vegetative growth.

19.16 WEED MANAGEMENT

Soil cultivation between the rows is carried out at an early stage, merely to control weeds. Weeds can be controlled mechanically, by hand, using a hoe, chemically, or by combining all these methods. Integrated weed management is also good practice in carrot for controlling weeds. Fluazifop-P-butyl chemical may be sprayed over the crop, as an early postemergence herbicide, for the control of many annual and perennial grasses. The dosage depends on the grass species and its stage of growth; young weeds are controlled with lower dosages than old ones. Flurochloridone herbicide is applied as a preemergent, as soon as possible after sowing, for the control of a wide range of broadleaf weeds. Its major disadvantage is its fairly long residual action of about 6 months, which can damage susceptible crops grown after the treated carrots. It is, thus, not a good option where vegetables are grown in quick succession. Haloxyfop-R-methyl ester may be used as a postemergent for the control of annual and perennial grasses. Soil should be hoed frequently to allow proper aeration and to prevent discoloration of crown.

19.17 INSECT-PEST AND DISEASE MANAGEMENT

Aphids sometimes occur on the leaves and crowns, and flower stems of carrots. They suck the sap from the plants resulting in retarded growth, yellowing, and restricted seed production. Control can be achieved by spraying with a registered pesticide. Red spider mite is generally not a serious pest in carrots but the numbers can increase rapidly as it gets warm. Cutworms can cause problems throughout growth. They cut young seedlings after emergence and feed on or around the carrots shoulders later in growth. Cutworms can be controlled by applying registered pesticides. *Alternaria* blight is dark brown to black spots, some with a yellow edge, appearing on the leaves. The oldest leaves are more susceptible than younger ones. The disease can be controlled by disinfecting seeds with a seed dressing, containing Thiram, and sowing certified seeds. In areas where blight is a problem, carrots should not be cultivated on fields that remain fallow or crop rotation. Bacterial blight disease is characterized by brown spots developing on the leaves and brown stripes on the petioles. In seed crops, the flower stems and inflorescence can be affected, whereas in carrots, brown, horizontal lesions appear on the leaves.

19.18 HARVESTING

The carrot takes about 3 months to be ready for harvest. Harvesting can start after last irrigation by pulling by hand, a spade, or a country plow. Carrots are harvested when they have reached a diameter of 20 mm and more, still young and tender. Usually when the carrots have reached the mature stage, their base tips appear on the soil surface.

19.19 YIELD

Carrot is the most quickly growing fodder crop. The average yield of carrot fodder is 500–600 q/ha under good management conditions.

19.20 SEED PRODUCTION

Carrot requires cooler climate for its seed production. For seed production, entire leaves are harvested with a height of 5–6 cm in the months of January–February. Then the bottom one third portions of the root are chopped off. The top portions of the roots with trimmed leaves are then planted in a well prepared field. The spacing is maintained in 50–60 cm row-to-row and 25–30 cm plant to plant. Carrot is a cross-pollinated crop due to protandry and pollination is done by honey bees. Being a cross-pollinated crop, allow an isolation distance of 1000 m from other varieties. Opening of umbel starts from periphery and is completed within 6–7 days. Then the crop is required frequent light irrigation every 3–4 weeks. The flowers come in the month of April and are harvested in May. The average seed yield may vary from 6 to 8 q/ha.

19.21 NUTRITIVE VALUE

Deep purple and yellow to orange colored carrots are rich in carotene which is precursor of vitamin A. The tuber dry matter contains about 12–14% of which 1% crude protein, 9% carbohydrate, and 1% fiber. The carrot leaves are quite rich in calcium, phosphorus, and minerals.

19.22 UTILIZATION

Carrot is used as a fodder for livestock worldwide for its balanced nutrition. Roots are used for making soups, curries, pies, pickles, and for salad purposes. Sweet preparation "gajar halwa" prepared out of carrot is delicious and popular. Roots are also canned. Carrot roots are rich sources of α- and β-carotenes (1890 μg/100 g) and contain sucrose 10 times that of glucose or fructose. Carrot leaves are a good source of leaf protein. It is used as a fodder and for preparation of poultry feeds. Carrot has many medicinal properties. It increases quantity of urine and helps in elimination of uric acid. It has cooling effect and is beneficial for people suffering from gall stones, constipation, and heat troubles. Purple and black carrots are used for preparation of a beverage called "kanji" which is a good appetizer. In France, essential oil separated from seeds is used for flavoring liquors and all kinds of food substitutes.

KEYWORDS

- **carrot**
- **agronomic management**
- **fodder yield**
- **nutritive value**
- **utilization**

REFERENCES

Chatterjee, B. N.; Das, P. K. *Forage Crop Production Principles and Practices*; Oxford & IBH Publishing Co. Pvt. Ltd.: Kolkata, 1989.

Oregon State University (OSU). Carrots. Commercial Vegetable Production Guides, 2004. http://hort-devel-nwrec.hort.oregonstate.edu/carrot.html (accessed Nov 10, 2017).

Sorensen, E. *Crop Profile for Carrots in Washington State*; Washington State University Extension, 2000. http:// www.ipmcenters.org/cropprofiles/docs/wacarrot.html (accessed Nov 12, 2017).

CHAPTER 20

AMARANTHUS (PIGWEED)

G. C. BORA*

Department of Plant Breeding & Genetics,
Assam Agricultural University, Jorhat 785013, Assam, India

**E-mail: gobin_bora@yahoo.co.in*

ABSTRACT

Amaranthus can be grown as an important forage crop. This plant may be fed to cattle and pigs as forage and fodder. When supplied in moderation, it is regarded as an exceptionally nutritious fodder. However, excess amount may cause many health problems in domestic animals. The amaranth is grown throughout the year in India because of its low production cost and high yield. The amaranth is really unique in many respects. It is easy to cultivate in a kitchen garden or on large scale and responds very favorably to fertilizer and organic manure. Such forage may cause fatal nephrotoxicity, presumably because of its high oxalate content. Other symptoms, such as bloat, might reflect its high nitrate content. Amaranth contains phytochemicals that may be antinutrient factors, such as polyphenols, saponins, tannins, and oxalates which are reduced in content and effect by cooking. The average yield of green leaves is about 74–94 q/ha. The average yield in "Shyamoli" and "Rodali" has been recorded to be 126 q/ha and 111 q/ha, respectively.

20.1 BOTANICAL CLASSIFICATION

Kingdom: Plantae
Order: Caryophyllales
Family: Amaranthaceae

Subfamily: Amaranthoideae
Genus: *Amaranthus*
Species: Spp.

20.2 BOTANICAL NAME

Amaranthus spp.

20.3 COMMON NAME

Amaranthus, pigweed, Tampala, Tassel Flower, Flaming Fountain, Fountain Plant, Joseph's Coat, Love-lies-bleeding, Molten Flower, Prince's Feather, and Summer Poinsettia.

20.4 INTRODUCTION

Amaranthus is one of the most common leafy vegetable crops grown in India. In Western countries, it is grown mostly for grain and ornamental purpose. Like many plant species, *Amaranthus* can be grown as an important forage crop. This plant may be fed to cattle and pigs as forage and fodder. When supplied in moderation, it is regarded as an exceptionally nutritious fodder. However, excess amount may cause many health problems in domestic animals. The amaranth is grown throughout the year in India because of its low production cost and high yield. It is easy to cultivate in a kitchen garden or on large scale and responds very favorably to fertilizer and organic manure. Such forage may cause fatal nephrotoxicity, presumably because of its high oxalate content. Other symptoms, such as bloat, might reflect its high nitrate content. Amaranth contains phytochemicals that may be antinutrient factors, such as polyphenols, saponins, tannins, and oxalates which are reduced in content and effect, by cooking. The average yield of green leaves is about 74–94 q/ha. The average yield in "Shyamoli" and "Rodali" has been recorded to be 126 q/ha and 111 q/ha, respectively.

20.5 DESCRIPTION

Amaranthus belongs to the family Amaranthaceae which comprises 65 genera and 850 species. The genus *Amaranthus* includes 50–60 species, the leaves of which are edible. These are most important leafy vegetables of the tropical countries, that is, South Asia, Southeast Asia, East Africa, Central Africa, West Africa, Ethiopia, the Pacific, and Far East. The important species of leafy amaranth are: *A. tricolor* L., *A. dubious* Mart. ex Thell., *A. lividus*, *A. blitum*, *A. tristis* L, *A. spinosus*, *A. viridis* L. and *A. graecizans* L. The most popular grain amaranth species are *A. hypochondriacus* L., *A. cruentus* L., and *A. caudatus*.

20.6 DISTRIBUTION

The green leafy amaranthus is said to be the native of India. The centers of diversity for *Amaranthus* species are Central and South America, India, and Southeast Asia. The secondary centers of diversity are West Africa and East Africa. A wide variation is reported to exist within each species in growth habit, disease resistance, caste, and quality, thus offering considerable scope for future breeding programs. In India, grain amaranthus is grown along with the whole length of the Himalayas from Kashmir to Bhutan on the South Indian hills. In the plains of India, particularly in Tamil Nadu, Karnataka, Kerala, Delhi, Odisha, West Bengal, and Assam, it is grown for leafy greens.

20.7 PLANT CHARACTERISTICS

The amaranth plant is a grain and green crop plant. The plant develops long flowers, which can be upright or trailing depending on the variety. The flowers are used to produce the amaranth grain, while the leaves can be used as amaranth greens. It is an annual or short-lived perennial plant. Some amaranth species are cultivated as leaf vegetables, pseudocereals, and ornamental plants. Most of the species from *Amaranthus* are summer annual weeds and are commonly referred to as pigweed. Catkin-like cymes of densely packed flowers grow in summer or autumn. Approximately, 60 species are recognized with inflorescences and foliage ranging from purple and red to green or gold. Members of this genus share many characteristics and uses with members of the closely related genus *Celosia*. Amaranthus leaves are oval, 2–4-in. long, starting out green or dark red and changing to bright yellow, orange, or florescent pink at the top (Fig. 20.1). The foliage of all varieties of *Amaranthus* is edible, highly nutritious, and is described as tasting like spinach.

Bisexual or unisexual, amaranthus flowers are typically very small and usually prickly with bristly perianth and bracts. Bracts subtend the flower with two scarious or membranous bractlets. The flower's androecium holds usually five stamens located opposite the sepals. Its stamens are generally united for part or all of the length into a membranous tube or crown-like structure, sometimes with tiny appendages between the anthers. Leaves on the amaranthus are alternate, simple, estipulate, and generally whole.

20.8 CLIMATIC REQUIREMENT

Amaranthus is a warm season crop adapted to the condition of hot, humid tropics, but is also suitable for temperate climate during summer. Like many vegetable crops, it needs at least 5 h of sunlight a day to do well. Although it could be grown in rainy season but it is highly susceptible to water logging. It belongs to C4 group of plants, species with efficient photosynthetic abilities that respond best to full sunlight. It has rapid, short growth cycles, high net assimilation rates, low CO_2 compensation point, and low transpiration coefficient (Nath, 1987).

20.9 SOIL AND ITS PREPARATION

It grows in every type of soil but the best crop is harvested from fertile loamy soils. The proper drainage system in the field is necessary, because this crop is susceptible to water logging. It can tolerate somewhat dry soil too. The best growing soil pH range is between 5.5 and 7.5, but some of the strains are successfully grown in soils with the pH up to 10. The soil should be brought to a fine tilth by plowing three or four times and leveling.

20.10 VARIETIES

While growing amaranth for food, it is best to select varieties of amaranth that work well as a food crop. If amaranth is to be grown as a grain, some amaranth varieties to consider include: *Amaranthus caudatus*, *Amaranthus cruentus*, *Amaranthus hypochondriacus*, and *Amaranthus retroflexus*. If it is to be grown as leafy greens, some amaranth varieties best suited to this include: *Amaranthus cruentus*, *Amaranthus blitum*, *Amaranthus dubius*, *Amaranthus tricolor*, and *Amaranthus viridis*. There are "green" and red or purple types of amaranthus both nutritionally similar but aesthetically preferred based on local demand and choice. The varieties "Co-1" (*A. dubius*), "Co-2" (*A. tricolor* Linn.), "Co-3" (*A. tristis* Linn.), "Chotti Chaulai" (*A. blitum* Linn.), and "Badi Chaulai" (*A. tricolor* Linn.) are green types. The Kerala Agricultural University has evolved "Kannara Local" (*A. tricolor* Linn.) which is a red type. Recently Assam Agricultural University (AAU) has developed and recommended two amaranth varieties, namely, Shyamoli (JORAM-1) and Rodali (JORAM-2) for the state of Assam (Bora, 2016; Bora et al., 2013). The plant and leaf color of the first one is green whereas the second one is purple. There are once-over harvest types and many cut types of amaranth. Many varieties recommended by different states are presented in Table 20.1.

TABLE 20.1 Recommended Varieties of *Amaranthus*.

Sr. No.	Variety	Features
1	Pusa Kirti	Leaves green with broad ovate lamina, 6–8-cm long with 4–5-cm wide. Petiole 3–4-cm long. The stem green and tender. Suitable for growing in spring–summer (March–June) season in the plains. Maturity 30–20 days.
2	Pusa Kiran	Leaves glossy green with broad ovate lamina. The lamina 7–9-cm long with 6–7-cm wide. Petiole 5–6.5-cm long. The stem glossy–green. Suitable for growing in kharif (July–October) season in the plains. Maturity 25–30 days.
3	Pusa Lal Chaulai	The upper surface of leaf deep red or magenta, lower surfaces purplish red. The lamina 8.5-cm long with 6.5-cm wide. Petiole 4.2-cm long. The stem deep red in color. Suitable for sowing in both spring–summer and kharif seasons in the plains. Maturity 25–30 days.
4	Co-1	Dark green in color grows vigorously, high yielder. Released by TNAU, Coimbatore.

TABLE 20.1 *(Continued)*

Sr. No.	Variety	Features
5	Co-2	Dark green lanceolate leaves, 20–25 days duration, 10–11 t/ha. Developed by TNAU, Coimbatore.
6	Co-3	Green leaves, 20–25 days for harvest up to 3 months, multicutting, and yield 10–12 t/ha.
7	Co-4	A dual purpose type for greens and grains. Plants dwarf, 20–25 days for leaves, 80–90 days for grains. Yield 7–8 t/ha (green) and 2–2.5 t/ha (grain).
8	Lal Sag	Grown popularly in many states. Belongs to *A. mangostanus*, high yielding Indian variety. Produces seed early, has small flowers.
9	Arka Suguna	It is a pure line selection from an exotic collection from Taiwan (IIHR-47). Leaves are light green, broad, and succulent stem. First harvest in 25–30 days after sowing and 5–6 cuts in 90 days. It is moderately resistant to white rust. It is recommended for Karnataka.
10	Arka Arunima	A new mutant which is resistant to white rust and rich in calcium and iron and low in antinutrient factors like oxalate and nitrates. It is a multicut, purple amaranth variety. Its leaves are broad and dark purple in color. It becomes ready for first cutting in 30 days after sowing. Two subsequent cuttings can be taken at 10–12 days interval. It grows well in kharif and rabi–summer seasons.
11	Badi Chaulai	Green large leaves. Plants tall, 2–3 cuttings, suitable for summer and kharif sowing. Developed by IARI.
12	Chotti Chaulai	Green, plants dwarf, suitable for spring–summer season, 6–7 cuttings. Developed by IARI.
13	Amt 105	Green leaves. Stem green with reddish base. Suitable for central zone of Kerala, yield 13 t/ha.
14	Amt 237	Green leaves and stem. Suitable for Kerala, yield 15–16 t/ha (greens).
15	Shyamoli	Green leaves and stems good in taste. Suitable for spring–summer season. Yield 13 t/ha (greens). Duration 30–20 days for first harvest. Developed by AAU, Jorhat and recommended for the state of Assam.
16	Rodali	Purple leaves and stem, good in taste. Suitable for spring–summer season. Yield 11 t/ha (greens). Duration 30–20 days for first harvest. Developed by AAU, Jorhat and recommended for the state of Assam.

AAU, Assam Agriculture University; IARI, Indian Agriculture Research Institute; TNAU, Tamil Nadu Agriculture University.

20.11 SOWING TIME

In the plains of North and Northeast India it is normally sown in March–April to take as a summer crop. The rainy season crop is sown in June–July or at the break of monsoon. In southern states, where the climate is favorable, it is sown throughout the year.

20.12 SEED TREATMENT AND INOCULATION

The seeds of amaranth may be treated before sowing with fungicides like Thiram or emisan or captan @ 2 g/kg of seeds as a precaution against damping off disease which is common in warm nursery. The nursery bed may be drenched with 0.2% solution of captan. The seeds may be inoculated with biofertilizers like *Azospirillum* and phosphorus solubilizing bacteria (PSB) for making nitrogen and phosphorus easily available to the seedlings and plants. They will further enhance the activity of the microorganisms in the soil.

20.13 SEED RATE AND SOWING METHOD

In Kerala and Tamil Nadu, amaranth is a transplanted crop. It is directly sown in North India and Assam. The seed requirement for direct sowing is 2–3 kg/ha and 1 kg/ha for transplanted crop. The seeds of amaranth are very small and sown shallow at the depth of 1 cm in rows about 20–30 cm apart in flat beds, if repeated cuttings are to be made. The seeds are usually broadcast in flatbed and then transplanted after 3–4 weeks of sowing. If plants are to be uprooted, 10 cm spacing is sufficient. Germination is seen in 10–14 days at 70°F temperature.

20.14 CROP MIXTURE

The grain types are usually grown as mixed crops along with cereals, pulses, and vegetables.

20.15 NUTRIENT MANAGEMENT

It grows on residual fertility of the previous crop taken in the field. However, the basal application of 25–30 t of Farmyard manure per hectare at the time of field preparation ensures a good crop. Some of the fertilizer recommendations are as follows: 50 kg nitrogen, 50 kg phosphorus, and 20 kg of potash per hectare. Three top dressings of ammonium sulfate can be done, first 1 month after sowing and the subsequent dressings soon after the first and second cuttings. The clipping type amaranth "Co-3" requires a higher fertilizer dose 75 kg N, 25 kg each of P_2O_5 and K_2O per hectare. Amaranth is highly adapted and quite efficient at extracting necessary minerals from a poor soil. Nevertheless, it has a high potassium requirement.

20.16 WATER MANAGEMENT

Water is the most essential commodity for growth and development of vegetable crops. It is absorbed by roots and along with nutrients it is translocated to different parts of the plant. Water influences photosynthesis, respiration, transpiration, utilization of mineral nutrients, and cell division in plants. Vegetable crops require water almost throughout the growing season. However, water use is critical in certain specific growth periods of the plant, varying with the crop species. Water available to the crops should neither be in excess nor shortage as both are harmful to the plants. Water stress or even excess of water during critical period will adversely affect the yield and quality of vegetable. The leafy crops require uniform soil moisture throughout their development. In amaranthus, moisture in soil is very necessary at the time of sowing in heavy soils, and in light soils a light irrigation should be given soon after sowing if soil moisture is insufficient. During summer, it is necessary to irrigate the crop at 3–5 days interval. The flood irrigation is practiced in crops grown in rows or by broadcast on well-leveled land with heavy soil. Furrow system of irrigation may be practiced if the crop is planted on ridges or raised beds (about 15–22-cm high). Sprinkler irrigation is useful in sandy or sandy loam soil. In order to reduce the water loss drip irrigation may be practiced.

20.17 WEED MANAGEMENT

The crop of amaranth is erect in growth allowing enough of weed growth which may be removed from time to time. Hoeing is easy in the plots with row sowing. First weeding should be done at 20 days after seedling emergence.

20.18 INSECT-PEST AND DISEASE MANAGEMENT

Leaf webber is a serious pest of amaranth. It can be controlled by destroying the larvae manually. Ants are also causing attack of this crop which can be controlled by applying malathion dust in the soil. It is advisable to avoid application of insecticides in amaranth. To control seed-borne diseases, seed treatment with captan powder and to control soil-borne diseases, soil drenching with captan solution is advocated as stated above. Although diseases are not a severe problem, but leaf spot and white rust are two diseases which affect the amaranth plants. Leaf spots can be controlled by spraying Bordeaux mixture (5:5:50) or 0.3% Blitox three times at an interval of 15 days (Singh, 2009). White rust can be controlled by following crop rotation and by spraying Dithane M-45 or any other copper fungicide at the rate of 1 kg per 500 L of water.

20.19 HARVESTING

The crop becomes ready for harvesting about 3–4 weeks after sowing and subsequent cuttings may be done after 7–10 days. Cuttings may be done at 7 days interval in "Chotti Chaulai" and at 10 days interval in "Badi Chaulai". In both "Shyamoli" and "Rodali" which were developed by AAU, Jorhat and recommended for the state of Assam, cuttings may be done at 10 days interval provided that adequate moisture in the soil is maintained. The plants are harvested by cutting them periodically. The crop rejuvenates quickly after each cutting. It does not stand storage for more than a few hours under ordinary conditions.

20.20 YIELD

The average yield of green leaves is about 74–94 q/ha. The average yield in "Shyamoli" and "Rodali" has been recorded to be 126 and 111 q/ha, respectively.

20.21 SEED PRODUCTION

The agronomic practices normally followed for production of tender leaves and stems are also usually employed for seed production. In soil of high alkalinity, the quality of seed will be very much affected. For seed crop, a spacing of 30 cm × 30 cm is usually maintained. Amaranth responds well to fertilizer schedule of 50 kg N and P, and 30 kg K per hectare in respect of seed yield. In addition to the soil application, spraying of urea at the rate of 12 kg N/ha and micronutrients at the rate of 400 g/ha once in 10 days from first flowering increased the yield of good quality seed.

Since it is a cross-pollinated crop, an isolation distance of about 400 m has been recommended between two varieties for foundation seed and 200 m for certified seed (Anonymous, 1971). It is a quick growing crop and forms seed in about 10–12 weeks. This is advantageous because after taking a number of leaf cuttings, only few last cuttings can be omitted to produce seeds well in June. Rouging of the offtype plants should be done at different stages of crop growth. Harvesting of inflorescence starts from 20 days after flowering when the glumes turn brown in color and seeds turn black. Initial drying of the spikes is seen and drying the seeds to a 15% moisture content are usually practiced so that the seeds can be threshed with pliable bamboo sticks. Seeds are then sieved through 2-mm sieve. The seed yield is about 200 kg/ha and is slightly higher in "Badi Chaulai" and "Shyamoli."

20.22 NUTRITIVE VALUE

Amaranthus is widely adaptabable herb and distributed throughout the world. Nutritive value per 100 g of edible part of the plant is given in Table 20.2.

TABLE 20.2 Nutritive Value per 100 g of Edible Portion.

Moisture	85.7 g	Protein	4.0 g
Fat	0.5 g	Minerals	27 g
Fiber	1.0 g	Other carbohydrates	6.3 g
Calories	46	Calcium	397 g
Magnesium	247 mg	Oxalic acid	772 mg
Phosphorus	83 mg	Iron	25.5 mg
Sodium	230 mg	Potassium	341 mg
Copper	0.33 mg	Sulfur	61 mg
Chloride	88 mg	Vitamin A	9200 IU
Thiamine	0.03 mg	Riboflavin	0.10 mg
Nicotinic acid	1.0 mg	Vitamin C	99 mg

20.23 UTILIZATION

Amaranthus is used as edible greens, herbs, and grains in Africa, China, Greece, India, Nepal, Pakistan, and Tibet. The fresh tender leaves and stem give delicious preparation on cooking as in the case of other fresh leafy vegetables. Cooking is similar to spinach or spinach beet; it is a cheap vegetable for the common people and is highly rich in vitamin A and C. The leaves and whole plant is fed to the cattle and pigs as forage and fodder. Silage can be prepared by chopping and fed to the animals. Further, the chopped plants with rice bran and *chapar* are cooked in water and giving some amount of salt makes it tasty for consumption by the cattle. As it is having high amount of carbohydrate, the grains are used as an energy source in different food preparations. The amaranth plant is typically grown as a decorative flower in North America and Europe. The seeds are edible, raw or toasted, and can be ground into flour and used for bread, hot cereal, or as a thickener. In Kerala, it is used to prepare a popular dish known as *thoran* by combining the finely cut leaves with grated coconut, chili peppers, garlic, turmeric, and other ingredients.

20.24 SPECIAL FEATURES (TOXICITIES)

When supplied in moderation, it is regarded as an exceptionally nutritious fodder. However, excess amount may cause many health problems in domestic animals. Like many other species of *Amaranthus*, this plant may be harmful and even deadly when fed to cattle and pigs in large amounts over several days. Such forage may cause fatal nephrotoxicity, presumably because of its high oxalate content. Other symptoms, such as bloat, might reflect its high nitrate content. Amaranth contains phytochemicals that may be antinutrient factors, such as polyphenols, saponins, tannins, and oxalates which are reduced in content and effect, by cooking.

FIGURE 20.1 (See color insert.) Amaranthus in flowering stage.

KEYWORDS

- **amaranthus**
- **agronomic management**
- **fodder yield**
- **nutritive value**
- **utilization**

REFERENCES

Anonymous. *Indian Minimum Seed Certification Standards*; Central Seed Committee, Ministry of Foods and Agriculture, Community Development and Cooperation, New Delhi; 1971.

Bora, G. C. Pustikar Khadya Hisabe Morisha Sak. Published in Assamese in *Weekly Janambhumi*, Jorhat, Assam, March 16, 2016.

Bora, G. C.; Saikia, L.; Bhattacharyya, A.; Rahman, S.; Hazarika, G. N. "Shyamali" and "Rodali"—Two Promising Varieties of Vegetable Amaranth (*Amaranthus* spp.) Suitable for Cultivation in NE India. In *Book of Abstracts of the National Seminar*; Held at BNCA, AAU, Assam, 2013; Vol. 3, pp 20–19.

Nath, P. *Vegetables for the Tropical Region*; ICAR Low Priced Book Series No. 2; ICAR: New Delhi, 1987.

Singh, B. *Horticulture at a Glance*; Kalyani Publishers: Kolkata, India, 2009.

PART IV
Leguminous Perennial Trees

CHAPTER 21

SUBABUL (RIVER TAMARIND)

SAGARIKA BORAH[1], MOKIDUL ISLAM[2*], and T. SAMAJDAR[1]

[1]*Krishi Vigyan Kendra, ICAR Research Complex for NEH Region, West Garo Hills, Umroi Road, Umiam 793103, Tura, Meghalaya, India*

[2]*SMS, Krishi Vigyan Kendra, Ri-Bhoi, ICAR Research Complex for NEH Region, Umroi Road, Umiam 793103, Meghalaya, India*

[*]*Corresponding author. E-mail: mislam01d@yahoo.co.in*

ABSTRACT

Subabul is an ideal plantation crop due to its multipurpose use and so can be used by farmers for agroforestry. Subabul cultivation would be particularly appropriate for marginal farmers who do not have access to large land holdings for fodder production. Subabul is a small, fast-growing tree native to Southern Mexico and northern Central America, but is now naturalized throughout the tropics, known as subabul in India. It is used for a variety of purposes, such as firewood, fiber, and livestock fodder. It is also important to stabilize the exposed and degraded soils and to help regenerate natural succession. Dry matter productivity of subabul varies with soil fertility and rainfall. Edible forage yields range from 30 to 300 q dry matter/ha/year. Subabul is one the highest quality and most palatable fodder trees. It comprises digestibility of 55–70%, crude protein 20–25%, nitrogen 3–4.5%, ether extract 6%, ash 6–10%, N-free extract 30–50%, Ca 0.8–1.9%, and P 0.23–0.27%.

21.1 BOTANICAL CLASSIFICATION

Kingdom: Plantae
Orders: Fabales

21.6.1 HAWAIIAN TYPE

The plants are short, bushy, and remarkably drought tolerant. It is suited to hilly terrains in drought prone areas. It is a prolific seed producer and is good for fodder purpose. K-341 is a Hawaiian variety.

21.6.2 SALVADOR TYPE

It is tall, tree-like, and fast-growing, having maximum annual biomass production; and possesses large leaves, pods, and seeds than Hawaiian types; responds to high fertilization. Variety K-8 is useful for fodder which has variable length up to 35 cm, with a large gland (up to 5 mm) at the base of the petiole. The tree bears numerous flowers in globose heads. The pods are 14–26 cm × 1.5–2 cm and brown at maturity. Each pod bears 18–22 seeds.

21.6.3 PERU

Tall and extensively branching type and is ideal for fodder purpose.

21.6.4 CUNNINGHAM

It is a cross between salvador and peru types.

21.7 SOIL REQUIREMENTS

It cannot withstand water logging. It requires deep well drained neutral soil and can tolerate saline and acid soil. It can also be grown in steep slopes, hilly terrains, gravelly areas and sandy loams. It can grow under a wide range of conditions as a range plant, roadside plant, in pastures, etc. The land should, however, be cleared of bushes, plowed and leveled before sowing for better performance. The tree grows very well under alkaline soils and also performs under dry clayey soils. The growth is average under sandy, acidic and dry gravel soils. The performance is poor under marshy, usar lands and in high altitudes. Due to its long and strong taproot, subabul can penetrate deep in compacted soils. The species has foliage which fertilizes the soil, as the fallen leaves decompose fast and form good humus to add soil nutrients. The species is an excellent nitrogen fixer thereby helps to augment the soil fertility.

21.8 CLIMATIC REQUIREMENTS

Subabul is best suited for warm regions and grows well between 25°C and 30°C (Brewbaker et al., 1985) in regions of 500–2000 mm annual rainfall. Because of its strong and deep root system, the tree is highly drought resistant and tolerates dry season extending 8–10 months, but the productivity is reduced. It is restricted to elevations below 500 m but withstands variations in rainfall, sunlight, windstorm, slight frost, and drought. Heavy frost kills the plant, whereas light only defoliates the tree. It is a light demander and grows slowly under the shade, though tolerates partial shade.

21.9 SEED COLLECTION

Seeds come out of pods which grow in clusters, from mostly self pollinated flowers, which look like fluffy white ball. The seed has a waxy white coat, and needs to be treated. In a kg, giant types have about 20,000 seeds. In general there are 30,000 seeds in a kg of 100% purity with about 6% moisture. Ripe pods should be collected before they split and dried in the sun for 3–4 days. The pods then split, when seeds can be gathered by sieving. Seeds are viable for 3–4 years.

21.10 PRETREATMENT

As the seed coat is hard, they need pretreatment before sowing: This can be done with any of the following ways:

- Soaking the seed in hot water (80°C) for 2–3 min or in cold water for 3–4 days.
- Seeds can also be dipped in concentrated sulfuric acid for 4 min and then washed.
- The seeds should be sundried for about 1 h before sowing.
- Germination occurs in 7 days with 70–80% germinating.

21.11 PLANTING

Direct sowing of seed during monsoons gives good result, but the plants grow slower than nursery-raised seedlings and if there is prolonged drought, the germinated seed may dry up. A seed rate of 3–4 kg/ha is recommended. Sowing is preferably done during February–March in a nursery or in polythene bags or in situ at 2–3-cm depth. Seedlings (1.5–3 months old with six to eight leaves) are planted in the main field. A spacing of 1 m × 0.1 m is recommended for a pure crop of fodder, 1.5 m × 0.2 m for planting in boundaries and borders of coconut gardens, and 2 m × 0.2 m when raised along boundaries. Planting of seedlings can be done with the onset of rains in May–June or September–October.

21.12 SPACING

The most common spacing adopted are 1.27 m × 1.27 m (50″ × 50″) (i.e., 6200 plants/ha); 2 m × 2 m (2500 plants/ha); 3 m × 1.5 m (2222 plants/ha). However, the recommended spacing is 1.5 m × 1.5 m (4445 plants/ha).

21.13 AFTER CARE

Weeds are a major cause of failure or slow establishment. Regular weeding, till plants are 1–2 m tall, gives best results. Weeding with soil working round the plants to a radius of 0.5 m should be done at least thrice in the first year, and as many times as needed in the subsequent years.

21.14 PESTS AND DISEASES

Subabul generally has been free of serious insect and diseases, but is susceptible to jumping plant lice (psyllids) which have caused serious defoliation and mortality in some areas. Some varieties are susceptible to gummosis, which is most likely caused by *Fusarium* or *Phytophthora* species. Leaf spot fungus also can cause defoliation under wet conditions.

21.15 YIELD AND ROTATION

Dry matter productivity of *Leucaena* varies with soil fertility and rainfall. Edible forage yields range from 30 to 300 q dry matter/ha/year. Deep fertile soils receiving greater than 1500 mm of well-distributed rainfall produce the largest quantities of quality fodder. Yields in the subtropics, where temperature limitations reduce growth rates, may be only 15–100 q of edible fodder/ha/year (Brewbaker et al., 1985).

21.16 NUTRITIVE VALUE

Subabul is one the highest quality and most palatable fodder trees. It comprises digestibility of 55–70%, crude protein 20–25%, nitrogen (N) 3–4.5%, ether extract 6%, ash 6–10%, N-free extract 30–50%, Ca 0.8–1.9%, and P 0.23–0.27%.

The foliage is highly palatable to most grazing animals, especially compared to other forage tree. In addition, it is very persistent over several decades of cutting or grazing, is highly productive, recovers quickly from defoliation, combines well with companion grasses, and can be grazed with minimal losses from trampling or grazing.

Mimosine is metabolized by ruminants to goitrogenic 3-hydroxy-4(1H) pyridone (DHP) in the rumen but in some geographical areas, ruminants lack the organisms (such as *Synergistes jonesii*) that can degrade DHP. In such cases, toxicity problems from ingestion of *Leucaena* have sometimes been overcome by infusing susceptible animals with rumen fluid from ruminants that possess such organisms, and more recently by inoculating cattle rumina with such organisms cultured in vitro. Such measures have facilitated *Leucaena* use for fodder in Australia and elsewhere.

21.17 VARIETIES

Fodder quality improves with decreasing seed numbers and breeders have developed some low seed producing varieties of subabul.

21.17.1 SUBABUL CO-1 (P)

This is a selection of variety Giant pill K-28 of subabul by Tamil Nadu Agricultural University (TNAU), Coimbatore and released in 1984 for Tamil Nadu state. The selection is high yielding (green leaf fodder 85 t/ha) with high protein and drought tolerance.

21.17.2 FD 1423

This is an introduction of subabul (*Leucaena diversifolia*) by TNAU, Coimbatore and released for the state of Tamil Nadu in 1999. The introduction is highly psyllid tolerant and is suitable for rainfed conditions. The green leaf yield is 55 t/ha.

21.18 SUBABUL BISCUITS

Forage, packed in pellets and cubes, is internationally marketed as animal feed. Subabul biscuits are produced from leaves to reduce wastage and transport cost. Production of subabul biscuits was initiated on a small scale in a few districts in Tamil Nadu. A factory has been setup in Sengipatti village in Thanjavur district to produce these biscuits. Subabul leaves are collected and grinded and mixed with molasses and bone meal to produce biscuits. *Prosopis* pods and rice bran are also added to enhance the nutritive value of the feed. The biscuits are more digestible than other feeds and animals tend to consume as much as 20% more of subabul biscuits.

Subabul biscuits are also preferred to subabul leaves. Trials with milch cattle and buffaloes show the biscuits enhance milk production. Both the milking period and the milk yield increased. The yield was higher in rural areas than in urban centers. In some urban centers, the milk yield increased by 8–10% and in rural areas the improvement ranged from 10% to 20%.

21.19 TOXICITIES

Subabul is one the highest quality and most palatable fodder trees of the tropics, often being described as the "alfalfa of the tropics." The leaf quality compares favorably with alfalfa or lucerne in feed value except for its higher tannin content and mimosine toxicity to nonruminants (Hammond, 1995). Concentrations in young leaf can be as high as 12% and the edible fraction commonly contains 4–6% mimosine. Livestock feed should not contain more than 20% of subabul, as the mimosine can cause hair loss and stomach problems. Leaves have a high nutritive value (high palatability, digestibility, intake, and crude protein content), resulting in 70–100% increase in animal live weight gain compared with feeding on pure grass pasture. Toxicity problems from ingestion of *Leucaena* have sometimes been overcome by infusing susceptible animals with rumen fluid from ruminants that possess such organisms (Allison et al., 1992).

FIGURE 21.1 (See color insert.) Subabul.

KEYWORDS

- **subabul**
- **fodder yield**
- **nutritive value**
- **subabul biscuits**
- **toxicities**

REFERENCES

Allison, M. J.; Mayberry, W. R.; Mcsweeney, C. S.; Stahl, D. A. *Synergistes jonesii*, gen. nov., sp. nov.: A Rumen Bacterium that Degrades Toxic Pyridinediols. *Syst. Appl. Microbiol.* **1992,** *15*, 522–529.

Brewbaker, J. L.; Hegde, N.; Hutton, E. M.; Jones, R. J.; Lowry, J. B.; Moog, F.; Van den Beldt, R. *Leucaena—Forage Production and Use*; NFTA: Hawaii, 1985; p 39.

Hammond, A. C. Leucaena Toxicosis and Its Control in Ruminants. *J. Anim. Sci.* **1995,** *73*, 1487–1492.

CHAPTER 22

GLIRICIDIA (QUICKSTICK)

GOLAM MOINUDDIN*

Assistant Professor, Regional Research Station, Bidhan Chandra Krishi Viswavidyalaya, Jhargram, West Bengal, India

**E-mail: moinuddin777@rediffmail.com*

ABSTRACT

Gliricidia is a highly palatable crop for the ruminants like cattle, goat etc. It is also found as a fence crop and used as supporting tree for black pepper, as well as shade tree for tea, coffee, cocoa plantations. It forms a broad canopy when full grown. This tree is an important source of green manure, fodder, fuel, etc. In many areas it is considered to be the second most important multipurpose legume tree. Annual leaf dry matter production of Gliricidia varies from 20 to 200 q/ha/year. The plant can yield up to 22 kg of seed per hectare. Leaf contains crude protein 20–40%, crude fiber 15%, ether extract 4.25%, nitrogen-free extract 51.65%, total acid 11.40%, digestable crude protein 14.90%, and total digestable nitrogen 69.20%. Dry matter digestibility varies from 60% to 65%. However, digestibility of low quality feed can be enhanced by adding with this legume.

22.1 BOTANICAL CLASSIFICATION

Kingdom: Plantae
Orders: Fabales
Family: Fabaceae
Subfamily: Papilionoideae
Genus: *Gliricidia*
Species: *sepium*

22.2 BOTANICAL NAME

Gliricidia sp.: *Gliricidia sepium*, *Gliricidia maculata*, *Gliricidia brenningii*.

22.3 COMMON NAME

Madre de cacao (South America), kakawate (Philippines).

22.4 INTRODUCTION

Gliricidia is a deep-rooted, fast-growing, medium-sized, and thornless legume tree. It is highly palatable crop for the ruminants such as cattle, goat, etc. It is also found as a fence crop and used as supporting tree for black pepper, as well as shade tree for tea, coffee, and cocoa plantations. It forms a broad canopy when full grown. This tree is an important source of green manure, fodder, fuel, etc. In many areas it is considered to be the second most important multipurpose legume tree (Daizy, 2007). Gliricidia is used as good forage crop in cut and carry system specifically for cattle and goat. Crushed fresh leaves can be used as poultice. Gliricidia leaves when used as feed, it can increase both weight and milk production of both large and small ruminants. Sometime boiled leaves are used as vegetables.

22.5 DESCRIPTION AND PLANT CHARACTERISTICS

Gliricidia is a medium-sized thornless trees. It can attain a height of 10–15 m with basal diameter of 50–70 cm. The stem surface is generally smooth but varies in color. Leaves are of 30-cm long and alternately opposite or sub-opposite in nature. Leaflets are 20 cm in length with 1–3 cm in width. Inflorescence is of compound raceme in nature. Flower's color varies from pink to tinged white with yellow spot of the base of the petals. Tender fruits are green in color and become light yellow-brown at matured stage. Fruits are of 10–18-cm long and 2-cm wide with 4–10 seeds. Seeds are light to dark brown in color with shiny in appearance. They take generally 1–5 years for flowering and fruiting.

22.6 ORIGIN AND DISTRIBUTION

This fodder tree has been supposed to be native to Mexico, Central America, and Northern South America and then introduced to various parts of Africa and Asia. In addition to its native range it is cultivated in many tropical and subtropical regions, including the Caribbean, India, and Southeast Asia (Hughes, 1987).

22.7 CLIMATIC REQUIREMENT

Gliricidia is generally grown in seasonally dry climate up to an altitude of 1200 m. For this reason it is deciduous in nature; however, under secured rainfall it may become evergreen. The mean temperature for growth is 20–30°C. Gliricidia can be grown in low rainfall areas of tropics having annual rainfall 800–1700 mm. It can tolerate seasonal drought with mean annual rainfall of less than 600 mm. This tree can also tolerate high wind velocity.

22.8 SOIL

It can be cultivated in a wide range of soil from sand to heavy clay. It can also thrive well in acid soil (pH 4.5–5), slightly alkaline soil, and calcareous soil. For good crop stand it likes well-drained soil. It is also suitable for cultivation in atoll environment. Gliricidia is not generally grown in areas of cool, wet, poor or compacted soil, very acidic soil (pH < 4.5), or alkaline soil (pH > 9).

22.9 SOWING TIME

Gliricidia when propagated through cuttings, cuttings are planted on bunds at 50-cm spacing during rainy season. Planting is done immediately after cuttings are separated from stem. When cultivation is done in steep slope the spacing is maintained at less than 20 cm. To obtain highest forage production, planting density of 4 trees/m^2 is maintained. Plants raised from cuttings grow better from those grown from seeds.

22.10 SEED TREATMENT

It should be treated with fungicide like carbendazim, Thiram, zairam, and captan or by *Trichoderma viride*.

22.11 SEED RATE AND SOWING METHOD

As Gliricidia is generally cut frequently for forage purpose, plants do not get chance to form seeds, then this crop is propagated through vegetative cuttings. For selecting cuttings, branch should be more than 7 cm in diameter and 30–100 cm in length and brownish-green in color. At first "V" shaped cut is made at both ends and then it is inserted into soil up to a depth of 20–50 cm depending upon the length of the cutting. When sowing is done through seeds, seeds are soaked in water overnight, then soaked seeds are sown in small plastic sachets filled with mixture of soil, sand, and FYM at 1:1:1 ratio. Watering is done regularly. Seedlings come out generally in 8–12 weeks. After 8 weeks, seedlings are double-spaced for more light penetration and aeration. Then seedlings are planted at desired spacing, the whole process generally takes 3–4 months.

22.12 INTERCROPPING

Gliricidia has a great potential to be used in intercropping as it fix atmospheric N_2 into soil. So when this crop is intercropped with other crops, it can boost up crop yields significantly without the needs of chemical fertilizers. This crop is generally intercropped with tea, coffee, vanilla, yam, etc.

22.13 WATER MANAGEMENT

Irrigation is not generally done except in very dry climate when annual rain falls below 600 mm. However, waterlogging situation should be avoided.

22.14 WEED MANAGEMENT

For herbicidal treatments, simazine @ 1 kg a.i./ha can be used as preemergence herbicide for control of grasses and broad leaf.

22.15 INSECT-PEST AND DISEASE MANAGEMENT

It is mostly insect-free as it has insect repellent property. However infestation of aphid (*Aphis craccivora*) during onset of rains causes blackening of leaf surface. Disease like little leaf cause by mycoplasma-like organism, chocolate brown leaf spot, etc. are found to happen. Colletotrichum sp. causes small, dark brown leaf spot. *Cladosporium* sp. can cause leaf defoliation and brown scab-like lesions on petioles and stems. Leaf scorch/scald and powdery black leaf spot are reported to occur.

22.16 HARVESTING

Gliricidia is a cross-pollinated crop. Flowering begins at the start of dry season (November–March). Ripening of pods depend upon the climatic condition and usually take 45–60 days. Harvesting should be done in proper time as pods are dehiscence in nature. Therefore, ripen pods are collected before they dehisce. If pods become over matured, seeds come out due to burst out of pods and seeds can disperse up to a distance of 40 m.

22.17 YIELD

Annual leaf dry matter production of Gliricidia varies from 20 to 200 q/ha/year. The plant can yield up to 22 kg of seed per hectare.

22.18 NUTRITIVE VALUE

Leaf contains crude protein 20–40%, crude fiber 15%, ether extract 4.25%, nitrogen-free extract 51.65%, total acid 11.40%, digestable crude protein 14.90%, and total digestable nitrogen 69.20%. Dry matter digestibility varies from 60% to 65%. However, digestibility of low-quality feed can be enhanced by adding with this legume.

22.19 UTILIZATION

Gliricidia is used as good forage crop in cut and carry system specifically for cattle and goat. Crushed fresh leaves can be used as poultice. Gliricidia leaves when used as feed, it can increase both weight and milk production of both large and small ruminants. Sometime boiled leaves are used as vegetables.

22.20 SPECIAL FEATURES

Due to some smell of leaves, sometimes animals seem to refuse to take it as feed, but its palatability can be enhanced by adding molasses or salt or by wilting for 12–24 h before feeding. The cultivation of Gliricidia is also confined in tropical areas. Sometime problem in pollination occurs during cultivation due to absence of sufficient pollinators.

22.21 COMPATIBILITY

New plants are susceptible to competition from grasses during the first year, it is a good competitor.

FIGURE 22.1 (See color insert.) Gliricidia.

KEYWORDS

- **Gliricidia**
- **agronomic management**
- **fodder yield**
- **nutritive value**

REFERENCES

Daizy, R. B. *Ecological Basis of Agroforestry;* CRC Press: Boca Raton, FL, 2007; p 44.

Hughes, C. E. Biological Considerations in Designing a Seed Collection Strategy for *Gliricidia sepium*. *Commonw. For. Rev.* **1987,** *66*, 31–48.

CHAPTER 23

KHEJRI (PROSOPIS)

RAJ KUMAR*, B. S. KHADDA, A. K. RAI, J. K. JADAV, and SHAKTI KHAJURIA

SMS, Krishi Vigyan Kendra Panchmahal (ICAR-Central Institute for Arid Horticulture), Godhra-Vadodara Highway, Vejalpur, Panchmahal 389340, Gujarat, India

**Corresponding author. E-mail: rajhortches@gmail.com*

ABSTRACT

Khejri is the state tree of Rajasthan, linked with the socioeconomic development of Indian Thar Desert. This tree is termed as "kalptaru" or the "king of desert" owing to its food, feed, and medicinal values. It is believed that the local name khejri has been derived from the name of a village Khejrali in Jodhpur district of Rajasthan. The species is very common in arid and semiarid zones of the country and it is a multipurpose tree because almost every part of the tree is utilized. The tender pods are eaten green or dried after boiling, locally called as sangri used in preparation of curries and pickles and ripe pods are called as *khokha*, sweet which contain 9–14% crude protein, 6–16% sugar, 1.0–3.4% reducing sugars, and 45–55% carbohydrate. The pods and leaves are favorite feed for animals especially goats, ship, camel, etc. Its wood (stem and branches) is favored for cooking and domestic heating, hard and reasonably durable; the wood has a variety of uses for house building, posts, tool handles, and boat frames. The *Prosopis cineraria* (L.) Druce is an important tree for the Thar Desert with hard climatic adaptation, represents one of the lifeline in desert habitat. This is a species represents all five-F, namely, Forest, Fiber, Fuel, Fodder and Food. This tree is also mythological important in local communities. High value of this species recognized as a State symbol (state tree of Rajasthan). *Sangri* is light green in color, straight, roundish flat, soft, sweet in taste, length range from 13.1 to 20.2 cm, and weight between 0.97 and 1.75 g. The ripe pods are 13.5–22.5 cm in length, 0.49–0.65 cm in width, and 1.65–2.15 g in weight with 11.15–16.85 seeds per pod. The weight of 100 seeds is 0.66 g. A 5-year-grafted plant yields a harvest of about 4.25 kg tender pods (*sangri*) and 6.25 kg fodder (*loong*) per year.

23.1 BOTANICAL CLASSIFICATION

Kingdom: Plantae
Orders: Fabales
Family: Fabaceae
Genus: *Prosopis*
Species: *cineraria* (L.) Druce
Binomial name: *Prosopis cineraria* (L.) Druce

23.2 BOTANICAL NAME

Prosopis cineraria (L.) Druce; Synonyms: *Prosopis spicigera* L.; *Prosopis spicata* Burm

23.3 COMMON NAME

It is also known as *Jand Jhind*, *Jhand* (Punjab); *Janti*, *Chonksa* (Delhi); *Banni*, *Perumbaj* (Karnataka); Sani Sumri (Gujarat); *Kandi* (Sindh); *Jambu*, *Perumbaj* (Tamil Nadu); and *Jamichettu* (Andhra Pradesh).

23.4 INTRODUCTION

Khejri Prosopis cineraria (L.) Druce is the state tree of Rajasthan, linked with the socioeconomic development of Indian Thar Desert. This tree is termed as "kalptaru" or the "king of desert" owing to its food, feed, and medicinal values. It is also known as *Jand Jhind*, *Jhand* (Punjab); *Janti*, *Chonksa* (Delhi); *Banni*, *Perumbaj* (Karnataka); Sani Sumri (Gujarat); *Kandi* (Sindh); *Jambu*, *Perumbaj* (Tamil Nadu); and

Jamichettu (Andhra Pradesh). It is believed that the local name khejri has been derived from the name of a village Khejrali in Jodhpur district of Rajasthan. The species is very common in arid and semiarid zones of the country and it is a multipurpose tree because of almost every parts of the tree is utilized, (Srivastava and Hetterington, 1991; Saroj et al., 2002). The tender pods are eaten green or dried after boiling; locally called as sangri used in preparation of curries and pickles (Khasgiwal et al., 1969) and ripe pods are called as *khokha*, sweet which contain 9–14% crude protein, 6–16% sugar (Bhiyama et al., 1964; Bhandari et al., 1978), 1.0–3.4% reducing sugars (Gupta et al., 1984) and 45–55% carbohydrate (Jatarsa and Paroda, 1981). The pods and leaves are favorite feed for animals especially goats, ship, camel, etc. Its wood (stem and branches) is favored for cooking and domestic heating (Mahoney, 1990), hard and reasonably durable; the wood has a variety of uses for house building, posts, tool handles, and boat frames. The *Prosopis cineraria* (L.) Druce is an important tree for the Thar Desert with hard climatic adaptation, represents one of the lifeline in desert habitat. This is a species represents all five-F, namely, Forest, Fiber, Fuel, Fodder, and Food. This tree is also mythological important in local communities. High value of this species recognized as a State symbol (state tree of Rajasthan).

23.5 BOTANICAL DESCRIPTION

P. cineraria is a small to medium size tree, ranging in height from 3 to 5 m. Leaves are bipinnate, with 7–14 leaflets on each of 1–3 pinne. Branches are thorned along the internodes. Flowers are small and creamy-yellow, and followed by seeds in pods. The tree is found in extremely arid conditions, with rainfall as low as 15 cm (5.9 in.) annually; but is indicative of the presence of a deep water table. As with some other *Prosopis* spp., *P. cineraria* have demonstrated a tolerance of highly alkaline and saline environments (Gupta, 2003). *Sangri* is light green in color, straight, roundish flat, soft, sweet in taste, length range from 13.1 to 20.2 cm and weight between 0.97 and 1.75 g. The ripe pods are 13.5–22.5 cm in length, 0.49–0.65 cm in width, and 1.65–2.15 g in weight with 11.15–16.85 seeds per pod. The weight of 100 seeds is 0.66 g. A 5-year-grafted plant yields a harvest of about 4.25 kg tender pods (*sangri*) and 6.25 kg fodder (*loong*) per year (Rani et al., 2013).

23.6 ORIGIN AND DISTRIBUTION

Khejri is a multipurpose small to medium size tree. This is a species represents all five-F, namely, Forest, Fiber, Fuel, Fodder and Food. There are no definite records about origin of khejri but Indian Thar Desert is considered as home of khejri. In the Vedic times, khejri wood was primarily employed to kindle the sacred fire for yagna. For this purpose, two pieces of the hardwood, that is, one of the khejri and other of the Peepal (*Ficus species*) were rubbed together to produce fire, during the Mahabharata period, Arjun along with Pandava hid his Gandiva in the hallow stem of khejri. In Valmiki Ramayana, khejri has been described along with other trees occurring on the hills at Panchvati. Lakshman used the branches of khejri trees as beams for the thatch at Panchwati, wherein he along with Lord Rama and his concert Sita, stayed during the 14 years of exile.

It is distributed in arid portions of Western Asia and the Indian Subcontinent, including Afghanistan, Iran, Arabia, India, Oman, Pakistan, Saudi Arabia, the United Arab Emirates, and Yemen. It is an established introduced species in parts of Southeast Asia, including Indonesia. In India, it is found in Rajasthan, Haryana, Delhi, Gujarat, Maharashtra, Madhya Pradesh, and North Karnataka. Khejri is a prominent constituent of desert thorn forest; pollen stratigraphic studies of Singh (1970) revealed abundance of khejri in western Rajasthan during the last 10,000 years when it also enjoyed a higher ecological status. The *Prosopis* trees are growing extensively under natural conditions in arid and semiarid regions of the country and distributed in all arid land forms, except the hills and saline depressions receiving 150–700 mm rainfall but plant density is better in areas receiving 350–500 mm annual rainfall. The densities of plant ranges from 5 trees/ha in western parts to 250 trees/ha in eastern parts. The highest densities occur in areas having 300–400 mm rainfall (Shankar, 1980).

23.7 PLANT CHARACTERISTICS

Khejri is a deep-rooted small to moderate to large sized tree, grows up to 10 m in height after 10–15 years and forms an open crown with thick, rough and gray bark, evergreen or nearby so, strong stem, rather slender branches armed with conical thorns with light bluish-green foliage. It does not ordinarily exceed a height of 12 m and a girth of 1.2 m, the maximum recorded being 18 m and 5.4 m, respectively. The leaves are 2.5–8.0-cm long, alternate, bipinnately compound with rachis 3–5-cm long, and glabrous; leaves are pinnate, mostly two pairs, and opposite; and 7–14 pairs of 4–15-mm long leaflets, and 8–23 pairs of 2–4-mm broad leaflets, sessile, oblique rounded, and mucronate at apex. Flowers are small, yellowish, auxiliary, and in slender spikes 7–11 cm in length, rachis 3–5-cm long, glabrous; pinnate mostly two pairs, opposite. Pods are cylindrical, straight, torulose and seeds are embedded in brown pulp, seeds 10–18, longitudinal ovate and dull brown. Flowering takes place during February–March and seeds mature during April–June. The seeds mature earlier in southern states than northern states: mid-April in Karnataka and Maharashtra, May in most parts of Gujarat and Rajasthan, and June in northern Rajasthan and southern Haryana. They are dispersed by wind and animals (Arya et al., 1992c).

Its flowers are entomophilous and depend on insects for seed setting. Pods are usually 10–20-cm long, rigid, cylindrical, glabrous, having thin exocarp, pulpy mesocarp, and papery endocarp containing 10–15 seeds. Phenology (Flowering time: February–March; Fruiting time: April–May; Leaf shedding: June–August; Leaf initiation: January–February; Pod maturity: June–July).The tree prefers a dry climate, and the most important areas of its distribution are characterized by extremes in temperature and found in dry and arid regions of India. Hence, it became an integral part of the traditional agriculture and the lifeline of the desert inhabitants.

23.8 SOIL

It can be grown in all types of soils except hills and saline depressions but mainly found in alluvial, coarse, and sandy, often alkaline soils where the pH may reach 9.8 (Arya et al., 1991, 1992a). It is a light demander and the older plants are drought resistant.

23.9 CLIMATIC REQUIREMENT

It grows luxuriantly under desert climatic conditions and makes excellent new foliar growth, flowering, and fruiting, where other desert plants are leafless and dormant during extremes of summer months (March–June).

It is very tough plant and tolerates both extremes of −4°C during winter and 50°C during summer (Arya et al., 1992c). It is commonly found under 75–500 mm annual rainfall.

23.10 GERMPLASM CONSERVATION

The wide range of genetic variability exists in arid and semiarid regions of country. Samadia et al. (2002) reported the same work carried by Central Institute for Arid Horticulture (CIAH), Bikaner, Central Arid Zone Research Institute, and AFRI, Jodhpur, Rajasthan and developed some technologies. So far, very little work has been undertaken for conservation either in situ or ex situ. The khejri tree has been traditionally protected by the *Bishnoi* community of Rajasthan, a remarkable phenomenon in the field of conservation of vegetation and soil.

23.11 VARIETAL WEALTH

The limited research work has been done for development of khejri variety and at present only one documented variety is released.

Samadia et al. (2002) reported 18 khejri genotypes bearing sweet (not acrid) pods were identified from a natural population of 600 trees in the vicinity of Bikaner area. The observations were made on pod characters such as length (5.34–27.43 cm), diameter (0.33–0.62 cm), weight (0.52–5.36 g), seeds per pod (6.81–27.40), yield (0.12–1.62 kg/m^3 tree canopy), and quality rating on 5-point scale. The significant differences in these characters showed the presence of sufficient variability in test trees. On the basis of overall performance, seven trees, that is, K1, K2, K4, K9, K11, K16, and K17 were found to be promising.

Thar Shobha: This variety released by CIAH in 2007. This is high-yielding and better quality variety. It has been recommended for uniform tender pod harvesting for vegetable use. A budded/grafted 5-year-old tree yields about 4.25 kg tender pods (*sangri*) and 6 kg dry fodder per year. With improved production technology biomass production is increasing with plant age. The tender pods at marketable stage are 13.1–20.2 cm in length, 0.17–0.42 cm in width, 0.97–1.75 g in weight, and seeds not separable in the stage of vegetable use. The tender pods are light in color, straight, roundish-flat, soft, and sweet. The ripen pods are 13.5–22.5 cm in length, 0.49–0.65 cm in width, 1.56–2.15 g in weight with 11.15–16.85 seeds/pod. The weight of 100 seeds is 6.66 g.

Recently, the research work was carried out at the CAZRI in Jodhpur, showing the potential that exists for the selection of fast-growing lines. This research would benefit greatly from the use of tissue culture technique in order to obtain large populations of uniform, high-yielding and fast-growing trees.

23.12 PROPAGATION

23.12.1 SEED PROPAGATION

Natural regeneration in khejri is difficult due to hard seed coat; hence it regenerates from root suckers (Arya, 1992a). The plants propagated by seed have extreme variability within population due to self-incompatibility (Felker et al., 1981). In *Prosopis species*, seed germination is a problem due to seed coat dormancy and the degree varies among species (Clements et al., 1977). The germination percent can be increased soaking seeds in water at room temperature for 48–72 h (Puri et al., 1992) and treating of seed with concentrate H_2SO_4 and HNO_3 for 5 min enhanced the seed germination. Seeds (25,000/kg) remain viable for decades in dry storage and establish well with 80–90% germination (Mahoney, 1990).

23.12.2 VEGETATIVE PROPAGATION

23.12.2.1 ROOT SUCKERS

Khejri is naturally regenerated by root suckers. However, the root suckers can be used for planting after separation from the mother plant but this is not an efficient technique for propagation of khejri because of availability of large number of suckers; and removal of suckers and planting has lot of mortality.

23.12.2.2 CUTTING

It can be propagated from stem cutting of juvenile. The plants developed through cutting are true to type or clone material (Arya et al., 1993).

Stem cutting of 15–20-cm long and 3–6 mm diameter, from 6-month-old seedlings were treated with auxins, applied singly or in combinations. The rooting frequencies were higher in February to May than August or November. The maximum 60% rooting was obtained with naphthalene acetic acid (NAA) +IBA +Thiamine 4+4+4 g/L during February. The IAA, IBA 2, 4-D and NAA, IBA, thiamine in concentration of 2000 and 4000 mg/L, respectively stimulated rooting up to 60% in stem cutting taken from 6-month-old plants, while the same combination of auxins with concentration of 1000 and 4000 mg/L, respectively induced rooting up to 35% in cutting from 8-year-old trees. Other treatments with auxin supplied singly (IAA, 2000 and 8000 mg/L; IBA, 2000 and 4000 mg/L; NAA, 1000 and 8000 mg/L) induced rooting in 10–50% of cuttings from 6-month-old plants. None of the treatments induced rooting from 8-year-old trees (Arya et al., 1994a).

23.12.2.3 AIR LAYERING

Khejri can be propagated by air layering. The 100% success in air layering reported on 40-year-old trees by application of Seradix B_3, Rootone and IAA 1000 ppm during July and August (Solanki et al. 1984, 1986). Under extreme hot arid conditions, air layering was not found as effective technique, but under good monsoon year some success was obtained.

23.12.2.4 BUDDING

Pareek and Purohit (2002) suggested that khejri could be successfully propagated by patch budding in the month of March with 76% success. The budded plants have very tiny and soft spines. Thus, in situ budding is suggested for block plantation of khejri in rainfed conditions.

23.12.2.5 MICROPROPAGATION

Recently, emphasis has also been given to propagate this species through micropropagation; but it appears that in vitro propagation is rather difficult with *P. cineraria* than with many other *Prosopis species*. The *Prosopis species* show a certain degree of recalcitrance to in vitro culture (Harris, 1992). Some work related to regeneration of *Prosopis species* through micropropagation has been done in India. In *Prosopis*, however, kinetin with IAA induced higher rate of shoot multiplication than BAP (Goyal and Arya, 1979, 1984). Shoot tips 1.5–2.0-cm long were excised from seedlings of each species 17–43 days after sowing (depending on growth rate) and placed on MS medium with added glucose, agar, glutamine, and IBA (1 mg/L). They were transferred to fresh medium after third and sixth week, and at eighth week to fresh medium containing 3 mg/L IBA and 0.01 mg/L kinetin. Very poor rooting was observed (Walton et al., 1990). A method for rapid in vitro propagation of *P. cineraria* through auxiliary branching is reported. Single node segments from actively growing branches of elite tree (4–5 years old) cultured on MS basal medium containing 3.0 mg/L each of 2-naphthoxy-acetic acid (NOA) and NAA produced auxiliary shoots. After 7–10 days, 2-cm-long shoots were transferred to modified MS medium containing 3 mg NOA/L. Within 25–30 days, 80% of shoots showed rooting as each shoot grew to a length of about 8 cm. The shoots were then cut into 5–7 segments and planted individually on MS + NOA. Each segment produced a plantlet. Following this procedure, 5–7-fold multiplication of plants could be achieved within a month; 30% of the plants survived after transplantation (Kackar et al., 1991).

23.12.2.6 ROOTSTOCK

There are some other important species in genus *Prosopis*, namely, *Prosopis juliflora*, *Prosopis alba*, *Prosopis Nigra*, etc, which can be utilized as rootstocks after ascertaining the graft compatibility and subsequent growth and fruiting. So far, *P. cineraria* is propagated on its own rootstock.

23.13 ORCHARD ESTABLISHMENT

Khejri is a tree and hence be planted at 8–10 m spacing. The budded plants being dwarf in stature can be planted at closer spacing (4 m × 4 m, 6 m × 6 m, 6 m × 4 m, 8 m × 6 m, and 8 m × 4 m) and accommodate 625, 277, 208, 416, 208, and 312 plants per hectare, respectively. The plants were planted in square and rectangular system of planting because of easy layout and intercropping during initial stage of orchard establishment. Mahoney (1990) suggested that it can be planted in close lines as a hedge with 1 m spacing between trees but tree densities of 50–100/ha are recommended for both agroforestry and silvi–pastoral systems. The field must be layout as per the spacing adopted. The pits of size 60 cm × 60 cm × 60 cm are dug during April–May and kept open for 15–20 days for soil solarization. The pits are filled with the well-rattan farmyard manure (FYM) and top pit soil in equal proportion. The 1-year-old plants are planted in the center of pit during monsoon; and irrigate the plants if there is no rain. Planting may be done during February with ensured irrigation available. The 9–12-month-old khejri plants are transplanted in pits of for better survival (Muthana et al., 1976). For successful establishment these are planted with onset of monsoon at 30–50 cm depth by removing polyethylene bag but without disturbing the block of soil and exposing the root; and irrigate the plant for the establishment of roots. Planting in pits or bored holes was found better, than planting by digging with a spade as in the latter case there was more loss of moisture. Gupta and Muthana (1985) reported higher survivability and better growth and establishment of seedlings of *Acacia tortilis* planted in the sandy plains. The previously made pits were used by placing pond sediments as subsurface moisture barrier at 60 cm depth in 5 mm thickness and refilled by mixing manure at the rate of 5 kg/pit. The higher seedling survivability and better growth and establishment are attributed to reduction in percolation loss and more availability of water to the young seedlings. In the absence of adequate rainfall post-planting irrigation helps in the establishment of seedlings. About 10 L of water per seedling per irrigation has generally been found adequate for the first 2 years. Soil mulching after irrigation helps in conserving moisture by reducing evaporation loss. The young saplings are sensitive to frost and drought but mature plants are tough and can resist drought. However, better option for establishment of khejri plantation especially in rainfed conditions is in situ where polybags raised 6–12-month-old seedlings are transplanted during the rainy season at desired spacing in already filled pits with proper filling mixture and budding should be done on these stocks in next season (March) by the scion buds taken from variety or promising types. Purohit et al. (2003) reported the growth vigor of khejri seedlings as affected by environment (in open and under shade net), irrigation method (rose can and minisprinkler), growth regulator treatment 50 ppm gibberellic acid (GA), 50 ppm GA + 2% urea, 1000 ppm maleic hydrazide (MH) and on the growth of the in situ budded plants as affected by black polyethylene mulch and application of 1 L urea (2%) per plant under irrigated and rainfed conditions. Seedling height after 6 months was 34.33 cm in plants treated with 50 ppm GA when planted under open with rose can irrigation compared to 29.6 cm in the control. Linear growth of budded plants was 123.31 cm under rainfed conditions when mulched with black polyethylene compared to 76.98 cm in the control.

23.14 NUTRIENT MANAGEMENT

So far, the farmers do not have much attention adding nutrient in natural grown plants. The very meager efforts have been made by researchers about nutrition management in khejri. Recently due to knowing the importance of this multipurpose tree the farmers are growing orchard systematically. The application of optimum dose of NPK at the seedlings stage has been done at Hissar and observed that N and K are critical for the better growth of this species at the seedling stage (Arya et al., 1991a). *P. cineraria* plants were grown first in sand (to height 10–12 cm), and then in pot culture in soil/farmyard manure (3:1) for 2–3 months and found that the species formed nodules which exhibited nitrogen-fixing activity (Pokhriyal et al., 1990). In contrast, the nodulation of 1–6-month-old progenies of 70 trees (10 from each of 7 districts of western Rajasthan), raised in the nursery were observed up to 5 months old, and only 0.7% of plants of all progenies were modulated at 6 months age, with no nodules occurring at all in 56 of the progenies. Nodules plant varied from 1 to 19 and nodule diameter from 0.01 to 0.4 cm (Kackar et al., 1990). Being a leguminous tree, it promotes the fertility of the soil under its canopy, supports denser, richer, and more productive ground vegetation (both annual and perennial species) than other dry land tree species (Srivastava and Hetherington, 1991). The improvement in soil organic matter content (Aggarwal et al., 1976; Lahiri, 1980) and contribution of higher soil moisture content (Gupta and Saxena, 1978); *P. cineraria* is well recognized. Puri et al. (1992) have also reported that nutrient levels (N, P, and K), moisture content, and organic carbon content of soil were higher under *P. cineraria* canopies than in the open.

23.15 WATER MANAGEMENT

The khejri trees grow naturally on sand dunes, wasteland, saline, alkaline, common land, forestland, and farmlands; but nowhere trees are irrigated.

However, the tree responds well to irrigation and it has been observed that the plants of khejri growing in farmland areas, where cropping is practiced, the growth of khejri plants are better than those plants growing on wastelands or community lands due to better moisture and nutrients availability, intercultural operations, protection from wild animals, etc., which is primarily done for crops. Moreover, plants are so tough that they can flourish well even under rainfed conditions but for initial establishment, little watering is required. In India, so far no work has been done on the water requirement of this valuable crop to improve is productivity. Under hot arid ecosystem, the plants are surviving because of its inherent morphological features like deep root system, spines, smaller leaves, corky bark, etc. In Oman, studies have been conducted for its ability to withstand high-air temperatures combined with low water availability. Plants were also sprayed with antitranspirant to reduce water loss. This limited the potential for evaporative leaf cooling and resulted in high leaf temperatures in khejri. Diurnal measurements were made of leaf and air temperature and of rates of photosynthesis and transpiration. Transpiration continued through the hottest periods of the day and appeared to be essential for leaf cooling. The species appeared to be impaired photosynthesis after exceeding threshold leaf temperature (Laurie et al., 1994). In a runoff harvesting and water conservation technique Gupta et al. (1995) observed, the overall best treatment was the 1.5 cm diameter saucer with weed removal and soil working; this improved soil moisture storage considerably, caused an increase in total biomass compared with the control (from 4.49 to 37.16 q/ha), doubled the root mass (from 4.33 to 9.66 q/ha), increased tree height by 20% and water use efficiency from 4.78 (in the control) to 39.6 kg/cm/ha. The interrow slopes technique was also effective. Gains in tree growth and water use efficiency in the other water harvesting treatments were relatively lower, although significant as compared to the control. Gupta (1984) reported construction of compacted circular catchment of 1.5 m radius with 5–10% slope around the transplanted plants. The result of field trials show that even during low rainfall years from 1986 to 1988, circular catchment technique increased the mean soil profile moisture storage by 10–30 mm/m profile and improved the growth and fruit yield of ber plants. Also, tree seedlings of *Acacia nilotica* (babul) and *P. cineraria* (khejri) were successfully established with this technique.

23.16 WEED MANAGEMENT

Under natural conditions, no marked adverse effect of vegetation growing in association with khejri on sand dunes has been observed. Sometimes, the *Laptadenia pyrotechnica*, *Calligonum polygonoides, A. nilotica*, and *Zizyphus jujube* growing with the khejri plants protect the young plants from wild animals. However, one or two weeding is necessary during the first year owing to slow initial growth rate.

23.17 INTERCROPPING

In arid areas the farmer has been traditionally protecting "Khejri," "Bordi," and "Babul" trees in their farmlands. He has a fine conviction that these trees besides conserving, add to the fertility and overall productivity of soil. These species are drought hardy and well-adapted to the climatic conditions of the desert. Satyanarayan (1964) and Saxena (1977) reported that "Khejri" forms climatic climax of western Rajasthan and dominate the alluvial flats while "Bordi" is one of the main codominant on the arid and semiarid zones. Gupta and Sharma (1992) reported 30–40 trees/ha in areas around Jodhpur and as high number as 104 trees/ha in Sikar region without any adverse effect on crop production. Mann and Saxena (1980) reported the effect of different densities of "Khejri" on the major crops grown in different habitats. They reported tree densities do not show much variation in crop yields. Even as high a tree density as 80 tree/ha showed an improvement in crop production in sandy plains. The increase in yield beside other factors could be due to buildup of soil fertility (Aggarwal and Lahiri, 1977). Khejri and "Bordi" trees, besides improving productivity, supply 20–30 kg and 2–3 kg air-dried leaves "Loong" and "Pala" as fodder, respectively (Saxena, 1984).

The crops like pearl millet, cowpea, cluster bean, toria mustard, and taramira can be raised successfully in the interspaces of well-grown khejri trees of 20 years age, as the fodder (green and dry) and grain yield of all the crops was more in association with trees as compared to sole cropping. The fodder as well as grain crops grown in khejri-based agri–silvicultural system earned more profit than sole roping. Higher profitability in agri–silvicultural system may be attributed to increased productivity of crops in this system and additional value of fuel and fodder from the associated khejri trees. In khejri-based agri–silvicultural system, the crop grown between the interspaces of trees showed more yields than sole cropping. Ahuja (1981) also observed better growth of annual crops under the trees than in open field. This may be due to less competition between trees and crops for moisture and light, as khejri is deep-rooted tree species with its monolayered tree canopy (Bisht and Toky, 1993). It may also be ascribed to improved soil fertility (Young, 1989) and ameliorative influence of shade in a hot–dry environment (Bunderson et al., 1990). Thus, in dry areas like Indian arid region, associated trees can reduce soil and water loss by alleviating understorey temperature and evapotranspiration (Belsky and Amundson, 1992) and may influence the productivity of ground vegetation.

In rabi, the crops were sown in the interspaces of khejri after lopping which removed hindrance for light. The higher yield of rabi crops in association with trees may be due to improved soil fertility (Young, 1989) and moisture retention, as a result of increased organic matter content of soil (Aggarwal et al., 1993). Kumar et al. (1992) also recorded higher yields from pearl millet, cluster bean and cowpea in association with khejri.

The fodder as well as grain crops grown in khejri-based agri–silviculture system earned more profit than sole cropping. Higher profitability in agri–silviculture system may be attributed to increased productivity of crops in this system and additional value of fuel and fodder from the associated khejri trees. Conservation of rainwater in kharif and subsequent rising of mustard in rabi season was found to be most economical practice. Similar results have been reported by Kaushik et al. (2000) in agri–silvi–horti system. More economic gain in khejri based agri–silviculture system, in comparison to sole cropping, has also been reported by Jaimini and Tikka (1998). The increased returns from tree–crop combination over the sole crops are supported by the studies of Reddy and Sudha (1989).

23.18 TRAINING AND PRUNING

Training and pruning are essential practices for providing strong structure, canopy management, and harvest of crop produce in khejri. After successful intake of scion bud on the rootstock, the upper portion of the scion should be removed so the growth of sprouted scion bud may faster. The wild sprouts bellow the graft should be removed time to time through pruning operations the grafts should be removed regularly through pruning operations using sharp cutter. To provide better framework of grafted tree, training operations should be done twice in a year like as November and June month and should be continued up to 5 year from budding. The regular training and pruning operations should be started after age of 5 years for better canopy management and yield potential of plantation keeping the objectives of crop production.

23.19 CROP REGULATION

So far, khejri is considered as forest trees and known primarily for its fodder value. Regarding sangri production and crop regulation very little efforts have been made. Villagers, however, collect sangri during the summer (April) at tender stage from the natural population. However, picking varies from place to place depending upon flowering and fruit maturity (early, mid, and late). The fruiting is also influenced by the management practices like time of lopping, agroecological conditions, influence of pest and diseases, etc. There are some plants where bearing twice in a year has also been observed. Since khejri is well prone to canopy management, therefore there is good possibility of crop regulation. In fact the trees which are lopped during the winter for loong production do not come in bearing during summer. This clearly indicates that the shoots require some time for physiological maturity for further bearing. Based on field observations, it is suggested that this valuable aspect should be given high priority for improving productivity of sangri.

Generally, the khejri trees are lopped (pruned) during November–December to harvest the leaf fodder (loong). These pruned trees do not flower as the new shoot sprouting after pruning are still immature during the flowering time of khejri in February–March and this do not produce sangria. When sangri is desired to be harvested from a particular tree, it is not lopped during November–December, but then the tree would not yield any loong; with technological recommendations systematic plantation would be developed for sangria production and these will have to be left unlopped. Consequently, these trees would not yield long. The studies on growth and fruiting regulation conducted at CIAH revealed that the plantation developed primarily for sangria production, the traditional practice of lopping during November–December could not be adopted, The pruning of plants at this time synchronize with the start of the second growth flush get sufficient time (6–7 months) to mature and differentiate floral buds which develop into flower and produce good crop of pods.

Thus, June pruning is recommended to harvest loong and sangria annually as a technique of crop regulation.

23.20 KHEJRI-BASED CROPPING SYSTEMS

Khejri is considered as a major component in various agri–silvi and silvi–pastoral systems in hot, arid region of northwestern India. The trees are also retained near the habitation as shade tree for providing shelter to man and animals. The villagers also organize various functions under the khejri trees. The crops such as wheat, maize, millet, sorghum, and mustard are generally grown with *Prosopis* (Arya et al., 1992a).The *Prosopis* and neem have close relationship and there is little competition for soil moisture and nutrients due to different root system (Arya et al., 1992b). The tree has a deep root system and can fix nitrogen (Arya et al., 1992a). Because of a deep taproot, trees do not compete for moisture or nutrients with crops growing close to the trunk. During the growing season it casts only light shade and is therefore suitable as an agroforestry species. Farmers in arid and semiarid regions of India and Pakistan have long believed it to increase soil fertility in crop fields. Yields of sorghum or millet increased when grown under *P. cineraria* as a result of higher organic matter content, total nitrogen, available phosphorus, soluble calcium, and lower pH (Mann and Shankarnarayan, 1980). Improvement in mechanical composition of soil (Singh and Lal, 1969), soil moisture content (Gupta and Saxena, 1978), physicochemical properties (Sharma and Gupta, 1989) and increase in dry matter yield (Aggrawal and Kumar, 1990) have also been reported under khejri tree.

In arid region of Haryana, chickpeas cv. C-235 was grown in a field without trees or in fields with scattered *P. cineraria* trees. The influence on crop growth of selected 15–18-year-old trees was determined. As the distance from the tree base increased, crop growth and yield initially increased (up to about 3.5 m) and then decreased. Beyond 14 m distance there was no effect on crop growth. Pod yield was highest on the north side of the trees. Soil contents of N, P, and K, soil moisture and organic carbon were higher under tree canopies than away from the influence of trees. *P. cineraria* had a positive effect on the growth of chickpeas (*Cicer arietinum*), with yields varying with direction and distance from trees (Puri et al., 1992). It was concluded that *P. cineraria* benefits chickpea growth and yield due to improvement of soil fertility and conservation of moisture (Puri et al., 1994). Sometimes it produces lateral roots in loosened soil, which compete for nutrient and water with the understory crops. Removal of lateral roots; however, does not affect the growth of the khejri adversely but enhances crop production under it (Yadav and Khanna, 1992). Vashishtha and Saroj (2000) reported that the vegetative vigor of ber, aonla, guava, bael, and pomegranate was better during initial years when they were grown with khejri as compared to sole plantation of these fruit crops. The multiple uses, prone to pruning, sparse foliage, deep root system, nitrogen-fixing capacity, high efficiency of recharging the soil with organic matter, etc. has made khejri trees a compatible component with different agroforestry systems.

Growth of natural pastures under *P. cineraria* is significantly higher (1.1–1.5 t/ha) than under *Acacia senegal* (0.6–0.7 t/ha). Maximum production of 2.6 t/ha has been obtained during 1973 with 27 rainy days and 641 mm of rainfall, clearly showing the advantage of a tree canopy in the semiarid conditions found in Rajasthan (Ahuja, 1980). *P. cineraria* are lopped for firewood and its productivity is greatly reduced when lopping is carried out every year; ideally, a rest of 4 years will result in 200% more leaf production (Saxena, 1980).

23.21 INSECT-PEST MANAGEMENT

23.21.1 BEETLES

The chaffer beetles (*Holotarchia* species, *Schizonycha fericollin* F., etc.) are the nocturnal feeders and main cause of defoliation of the trees. This can be controlled by spraying of carbaryl 50 WP at 0.15% at weekly intervals.

23.21.2 LOCUST

The damage of khejri plants by locust has also been reported by Bhanotor (1975). *P. cineraria* acted as a less suitable host for aphids and remained uninfested due to greater amount of sugar, lower amount of lipids, and the influence exerted by VAM fungi in imparting resistance (Murugesan et al., 1995).

23.21.3 SHOOT GALL MAKER

The stem and rachis galls of *P. cineraria* caused by *Lobopteromyia prosopidis* [*Contarinia prosopidis*] has been observed frequently. The phenolic contents (total phenols and O-dihydroxy phenols) of *L. prosopidis* [*C. prosopidis*]-induced gall tissue and healthy tissue of *P. cineraria* were studied in vivo and in vitro (Ramani and Kant, 1989). Removal of affected portion and spraying of insecticides like chloropyriphos (0.05%) at new flush stage are suggested for its control.

23.21.4 TERMITE

The termite is a major problem, which affect both old and new plants. Though, the damage due to termites was common in old trees. The colonies of termites originate at the base of trunk of older trees; later penetrate into trunk followed by extensive hollowing out of the heart wood. Soil drenching by chlorpyrifos and removal of termites colonies can minimize the problem.

23.22 DISEASE MANAGEMENT

Drying of khejri plants can be observed in many areas mostly in the canal irrigated areas, though the reason is yet to be investigated. Sharma et al. (1997) has reported root rot of *P. cineraria* caused by *Fusarium solania*s a new host in Haryana. The fungus *Ravenelia spiergerae* is a pathogen recorded in nursery. It can be managed by drenching with application of copper oxy chloride and Ridomil M Z.

23.23 MATURITY INDICES, HARVESTING, AND YIELD

The seedling plants of khejri come in bearing after eight years, while in situ budded plants starts bearing next year. Usually, khejri plants flowers during spring season, that is, February–March and tender pods are ready for harvesting in the month of April–May. Though the flowering and fruiting are influenced by the agroecological situations and management practices but in general pods are ready for harvest within 20 days of fruit setting. The green tender pods at papery soft seed stage can be harvested for sangri purposes, while the ripe pods can be harvested for preparation of cookies. The dried ripe pods are called *khokha.* The colored pod types khejri are also green in color at tender stage. Immature green pods with low tannin and fiber content and soft seeded are considered good for dehydration purpose (Nagaraja et al., 2003). Naturally the khejri tree are spiny, therefore it is very difficult to harvest the sangria. The normal method adopted in harvesting the pods are by climbing on the tree and the sangria plucked manually along with the some twigs and branches. The sangri sorted and inert materials are removed. The *khokha* can be collected from dropped out ripe pods also. The yield from tree depends upon bearing behavior, age of tree, and management practice. A good fruiting 15-year-old tree produces 10–15 kg sangri and 25–30 kg loong. Muthana (1980) has reported that the full grown tree (30–50 years) and unlopped tree produces about 5 kg of air-dried pods and about 2 kg seeds. Besides sangria, a 15-year-old tree produced 35–40 kg leaf fodder/trees in 350 mm rainfall area (Muthana, 1980). It produces 17.60 kg/tree biomass in at Hissar (350–400 mm rainfall) after 6 year (Toky and Bisht, 1992).

23.24 UTILIZATION

Khejri is a multipurpose tree whose all parts are used in various manners given below.

23.24.1 GREEN PODS

Green pods locally called "sangar" used to make dhal, the classic Indian dish. "Ker Sangri" is one of the most mouth-watering delicacies of Jaisalmer. Ker Sangri is a popular vegetarian dish of Jaisalmer. It is cooked on low heat. The ingredients that go into this dish are desert beans and capers. This vegetarian delicacy can be best enjoyed with "*Bajara roti.*" The *kandi* or *sagri pachadi* is easy to prepare and makes for a tasty chutney recipe also. *Kandi pachadi* or *tour dhal* chutney is served with rice, dosa, or idly.

Liu et al. (2012) reported dried pods of khejri are consumed as a vegetable, boiled with water to afford the aqueous extract. Extraction of the residue gave methanolic extract. Panwar et al. (2014) reported the unripe pods are also used as vegetable; boiled and dried pods are the important constituents of this region's famous dishes "Panchkutta" (ker, kumat, lasoda, amchur, and sangri). Meena et al. (2014) reported that several kind of traditional vegetables are consumed by the farming communities grown by the farmers in study area (Bikaner and Nagaur district) of the hot arid region of western Rajasthan. The majority of the farmers/local people of the study areas use the vegetables as fresh or after dehydration or value addition in their daily diet throughout the year. They convert them in form of value addition in different forms and use them as preserved fruits/vegetables as their daily dietary food stub.

23.24.2 RIPE PODS

Ripe pods locally called "*khokha*" are used fresh as well as preparation of bakery items such as biscuits and cookies. It contains protein (23.20%), carbohydrate (56.0%), fat (2.0), fiber (20.0), vitamin C (523.0/100 g), calcium (114.0 mg), phosphorus (400 mg), iron (19.0), and energy (334.8 kcal/100 g). The ripen pods with seeds are rich in protein and good source of animal feed (Rathore, 2009).

23.24.3 LEAVES FODDER

Leaves are most important top feed species providing nutritious and highly palatable green as well as dry fodder, which is readily eaten by camels, cattle, sheep, and goats, constituting a major feed requirement of desert livestock. Khejri trees are ready to provide animal feed from the 10th year onwards and, if properly managed, may be kept in production for 2 centuries. An average tree yields 25–30 kg of dry leaf forage per year. Dry matter intakes of 685 and 1306 g/day are quoted for sheep and goats, respectively.

P. cineraria have an important place in the economy of the Indian desert. In the arid zone of Rajasthan, camels, goats, donkeys, and mules, which make up about 40% of the 19 million head of livestock in the region, depend on browsing to meet their nutrient requirements. Khejri is well adapted to the very dry conditions in India and is found in zones with annual rainfall ranging from 150 to 500 mm the optimum density is seen between 350 and 400 mm. This plant produces its leaves, flowers, and fruit during the extreme dry months (March–June) when all other species adapted to arid zones are leafless and dormant. It is this characteristic which deserves greatest attention as the tree offers a new forage resource for extreme arid zones.

The importance of *P. cineraria* is well recognized by farmers as it provides an extra source of revenue, acts as an insurance against drought, and increases the sustainability of production systems in this drought-prone fragile ecosystem. Unripe pods (sangri)/ripe pods indicate that they can offer a good source of livestock feed compared to other native available feeds. Arid foods had are great nutritional values (Rani et al., 2013).

23.24.4 BARK

Bark is used as a local medicine to cure bronchitis, asthma, piles, lecucoderma, rheumatism, scorpion bites (Bhandari, 1974), and snake bite (Kirtikar and Basu, 1935). In order to avoid the abortion, the traditional healers of Chhattisgarh use the Herbal Mala prepared from *Shami* roots, though *Shami* trees are not common in Chhattisgarh. The traditional healers purchase the roots from local herb shops and use it. In general, the white-colored string is used to prepare herbal mala. The traditional healers advise their patients to use this herbal mala to wear it either around the neck or to tie it around their waist, in order to avoid abortion. Ground flower are used as a tonic, blood purifier, and curer of skin disease. Women use powdered flowers mixed with sugar during pregnancy to safeguard them against miscarriage. *Prosopis* gum also has the properties to those of *Acacia* gum (Arya et al., 1992a). Liu et al. (2012) reported leaves as traditional medicine to cure a wide range of diseases in the state of Rajasthan, India.

23.24.5 WOOD

Wood is reported to contain high calorific value and provide high-quality fuelwood. The lopped branches are good as fencing material.

23.24.6 GUM

Gum produces a brown shining gum just like Arabic Gum which is obtained during the months of April–June.

23.24.7 *FLOWER*

Flower is pounded, mixed with sugar, and used during pregnancy as safeguard against miscarriage.

23.24.8 *WORSHIP OF KHEJRI*

Worship of khejri during Vedic times, khejri wood was used to kindle the sacred fire for performing a yagna. In Hindu epics, the Ramayana and the Mahabharata, mention the usefulness and significance of this tree.

23.25 NUTRITIVE VALUE OF KHEJRI FRUITS (SANGRI)

Khejri fruits or pods are locally called sangri. The dried pods locally called khokha are eaten. Dried pods also form rich animal feed, which is liked by all livestock. Green pods also form rich animal feed, which is liked by drying the young boiled pods. The dried green sangri is used as a delicious dried vegetable which is very costly (Nearly Rs.800 per kg in market in the year 2015). Many Rajasthani families use the green and unripe pods (sangri) in preparation of curries and pickles. The nutritive value of sangria is protein (23.20 g), fiber (20.0 g), calcium (414 mg), phosphorus (400 mg), iron (19 mg), and vitamin C (523 mg).

23.26 NUTRITIVE VALUE OF LEAVES

The leaves are of high nutritive value; locally it is called "*loong*." Feeding of the leaves during winter when no other green fodder is generally available in rainfed areas is thus profitable. The pods are a sweetish pulp and are also used as fodder for livestock. The leaves of khejri are considered as excellent fodder in desert. The green leaves contain 14–18% crude protein, 13–22% crude fiber, about 6% ash, 44–59% nitrogen-free extract, 0.28–0.9% phosphorus, and 1.5–2.7% calcium. The leaves also contain 11.6% tannin on a dry matter basis on an average leaves contain 72 mg of sodium, 1.23 mg of potassium, 1.12 mg of iron 1.16 mg of zinc, 5.75 mg of manganese, and 1.87 mg of copper per 100 g of dry matter (Bohra and Gosh, 1980).

KEYWORDS

- **khejri**
- **agronomic management**
- **fodder yield**
- **nutritive value**
- **utilization**

REFERENCES

Aggrawal, R. K.; Kumar, P. Nitrogen Response to Pearl Millet (*Pennisetum typhoides*). **1990**.

Aggrawal, R. K.; Lahiri, A. N. Influence of Vegetation on the Status of the Organic Matter and Nitrogen of the Desert Soils. *Sci. Cult.* **1977,** *43*, 535.

Aggarwal, R. K.; Gupta, J. P.; Saxena, S. K.; Muthana, K. D. Studies on Soil Physio-chemical and Ecological Changes Under Twelve Years Old Five Desert Tree Species of Western Rajasthan. *Indian For.* **1976,** *102*, 863–872.

Aggarwal, R. K.; Kumar, P.; Raina, P. Nutrient Availability from Sandy Soils Underneath. *Prosopis cineraria* (Linn. Macbride) Compared to Adjacent Open Site in an Arid Environment. *Indian For*. **1993,** *199*, 323–325.

Ahuja, L. D. Grass Production under Khejri Tree. In: *Khejri in the Indian Desert: Its Role in Agroforestry*; Mann, H. S., Saxena, S. K. Eds.; Central Arid Zone Research Institute, Jodhpur, 1981; p 78.

Ahuja, L. D. Grass Production. In *Khajri* (*Prosopis cineraria*) *in the Indian Desert*: *Its Role in Agroforestry*; Mann, H. S., Saxena, S. K., Eds.; Monograph No, 11, CAZRI: Jodhpur, 1980; pp 28–30.

Arya, S.; Toky, O. P.; Bisht, R. P.; Tomar, R. *Prosopis cineraria* Promising Multipurpose Tree for Arid Lands. *Agrofor. Today* **1991,** *3* (4), 13.

Arya, S.; Bisht, R. P.; Tomar, R. Effect of NPK Fertilizers on Growth and Biomass Production of *Prosopis cineraria* (L) Mac Bridge seedling. *J. Agri. Biol. Res*. **1991a,** *7* (2), 112–115.

Arya, S.; Toky, O. P.; Bisht, R. P.; Tomar, R. India's Tree of Promise—*Prosopis cineraria. Farm For. Newslett*. **1992a,** *4* (3), 1.

Arya, S.; Toky, O. P.; Harris, P. J. C.; Harris, M. S. *Prosopis cineraria* and *Azadirachta indica* a Surprising Association. *Agrofor. Today* **1992b,** *4* (3), 9–10.

Arya, S.; Toky, O. P.; Bisht, R. P.; Tomar, R. *Potential of Prosopis cineraria* (L.) *Druce in Arid and Semi Arid India*. In Proceeding of the Symposium on *Prosopis cineraria* Aspects of Their Value, Research and Development CORD, University of Durham, UK, July 27–31, 1992c, 61–70.

Arya, S.; Toky, O. P.; Tomar, R.; Singh, L.; Harris, P. J. C. Seasonal Variation in Auxin-induced Rooting of *Prosopis cineraria* Stem Cuttings. *Int. Tree Crops J.* **1993,** *7*, 249–259.

Arya, S.; Tomar, R.; Toky, O. P. Effect of Plant Age and Auxin Treatment on Rooting Response in Stem Cuttings of *Prosopis cineraria. J. Arid Environ.* **1994a,** *27*, 99–103.

Belsky, A. J.; Amundson, R. G. Effect of Tree on Understorey Vegetation and Soil at Forest/Savanna Boundaries in East Africa. In *The Nature and Dynamics of Forest Savanna Boundaries*; Furley, P. A., Proctor, J. Eds.; Champion and Hall: London, UK, 1992.

Bhandari, M. M. Famine Foods of the Rajasthan Desert. *Eco. Bot.* **1974,** 28 (1), 74–75.

Bhandari, M. M. *Flora of the Indian Desert*; Scientific Publishers: Jodhpur, India, 1978.

Bhiyama, C. P.; Kaul, R. N.; Ganguli, B. N. Studies on Lopping Intensities of Prosop. **1964**, *5*.

Bisht, R. P.; Toky, O. P. Growth Pattern and Architectural Analysis of Nine Important Multipurpose Trees in an Arid Region of India. *Can. J. For. Res.* **1993,** *23*, 722–730.

Bohra, M. C.; Ghosh, P. K. The Nutritive Value and Digestibility of Loong. In *Khejri in the Indian Desert*; CAZRI: ICAR, 1980; pp 45–47.

Bunderson, W. T.; Wakeel, A. E. I.; Saad, Z.; Hashim, I. Agro Forestry Practices and Potential in Western Sudan. In *Planning for Agroforestry*; Budd, W., et al. Eds.; Elsevier Science Publisher: New York, 1990.

Clements, J.; Jones, R. G.; Golbert, N. H. Effect of Seed Treatments on Germination of *Acacia. Aust. J. Bot.* **1977,** *25*, 269–276.

Felker, R.; Cannei, G. H.; Clark, R. P. Variation in Growth Rate Among 13 *Prosopis cineraria. Exp. Agri.* **1981,** *17*, 209–238.

Goyal, Y.; Arya, H. C. Clonal Propagation of *Prosopis cineraria* Linn. Through Tissue and Differentiation. *J. Indian Bot. Soc.* **1979,** 58–61.

Goyal, Y.; Arya, H. C. Tissue Culture of Desert Trees: I. Clonal Multiplication of *Prosopis cineraria* a by Bud Culture. *J. Plant Physiol.* **1984,** *115*,183–189.

Gupta, J. P.; Saxena, S. K. Studies on the Monitoring of Their Dynamic of Moisture in the Soil and the Performance of Ground Flora under Desertic Communities of Trees. *Indian J. Ecol.* **1978,** *5*, 30–36.

Gupta, A. K.; Solanki, K.; Kackar, N. L. Variation for Quality of Pods in Prosop. **1984**, 3.

Gupta, G. N.; Bala, N.; Choudhary, K. R. Effect of Runoff Harvesting and Conservation Techniques on Growth and Biomass Production of *Prosopis cineraria. Indian For.* **1995,** *123* (8), 702–710.

Gupta, J. P.; Muthana, K. D. Effect of Integrated Moisture Conservation Technology on the Early Growth and Establishment of *Acacia tortilis* in the Indian Desert. *Indian For. III.* **1985,** (7), 477–4H5.

Gupta, J. P. Technology Approach for Greening of Degraded Arid Land Bulletin. 2003, 11.

Harris, P. J. C. Vegetative Propagation of Prosopis, In. *Prosopis Species: Aspects of Their Value, Research and Development*; Dutton, R., Ed.; Centre for Overseas Research and Development, University of Durham: Durham, UK, 1992; 175–191.

Jaimini, S. N.; Tikka, S. B. S.; Khejri (*Prosopis cineraria*) Based Agroforestry System for Dry Land Areas of North and North-west Gujarat. *Indian J. For.* **1998,** *23*, 331–332.

Jatarsa, D. S.; Paroda, R. S. *Prosopis cineraria* an Unexploited Treasure of the Thar. *J. Ecol.* **1981,** *5*, 30–36.

Kackar, N. L.; Jindal, S. K.; Solanki, K. R.; Singh, M. Nodulation in Seedlings of *Prosopis cineraria* (L.) Druce. *Nitrogen Fixing Tree Research Reports*. 1990; Vol. 8, pp 152–153.

Kackar, N. L.; Solanki, K. R.; Vyas, S. C.; Singh, M. Micropropagation of *Prosopis cineraria*. *Indian J. Exp. Biol.* **1991,** *29* (1), 65–67.

Kaushik, N.; Singh, J.; Kaushik, R. A. Performance of Arable Crops in the Initial Years Under Rainfed Agri–Silvi–Horticultural System in Haryana. *Range Manag. Agrofor.* **2000,** *23*, 170–174.

Khasgiwal, P. C.; Mishara, G. G.; Mitthal, B. M. Studies of *Prosopis spclgera gun.* Part I: Physicochemical Character. *Indian J. Pharm.* **1969,** *1*, 148–152.

Kirtikar, K. R.; Basu, B. D. *Indian Medicinal Plants*, 2nd ed.; Bishan Singh and Mahendrapal Singh Publications: Dehradun, India, 1935; Vol. II, pp 910–912.

Kumar, V.; Yadav, H. D.; Sharma, H. C. Agroforestry: The Suitable Farming System for Arid and Semi-arid Region; *Haryana Farming* 1992; Vol. 23, pp 15–16.

Lahiri, A. N. *Frasopis cineraria* in Relation to Soil Water and Other Conditions of its Habitat.1980.

Laurie, S.; Bradbury, M.; Stewart, G. R. Relationships Between Leaf Temperature, Compatible Solutes and Antitranspirant Treatment in Some Desert Plants. *Plant Sci. Limeric.* **1994,** *100* (2), 147–156.

Liu, Y.; Singh, D.; Nair, M. G. Pods of Khejri Consumed as a Vegetable Showed Functional Food Properties. *J. Funct. Food* **2012,** *4*, 116–123.

Mahoney, D. *Trees of Somalia: A Field Guide for Development Works*; Oxfam/HDRA: Oxford, 1990, pp 133–136.

Mann, H. S.; Saxena, S. K. Role of Khejri in Agroforestry. In *Khejri* (*Prosopis cineraria*) *in the Indian Desert: Its Role in Agroforestry*; Mann, H. S., Saxena, S. K., Eds.; 1980; pp 64–67.

Murugesan, S.; Sundararaj, R.; Mohan, V.; Mishra, R. N. Insect-VAM Interactions: Colonization of Aphid, Aphis nerii boy. (Homoptera: Aphididae) on Host and Non-host Trees. *Phytophaga* (*Madras*), **1995,** *7* (2), 123–126.

Muthana, K. D.; Arora, G. D.; Gian C. Comparative Performance of Indigenous Trees in Arid Zone Under Different Soil Working Techniques. *Ann. Arid Zone* **1976,** *15* (1,2), 67–76.

Muthana, K. D.; Sharma, S. K.; Raina, A; Meena, G. L. *Silvi-pastoral Studies*. Annual Progress Report, 1980, CAZRI: Jodhpur, 1980.

Nagaraja, A. Khejri Biscuits: A Viable Products from Khejri (*Prosopis cineraria*) Pods. CIAH, *News Letter* **2003,** *2* (1), 4–5

Panwar, D.; Pareek, K.; Bharti, C. S. Unripe Pods of *Prosopis cineraria* Used as a Vegetable (Sangri) in Shekhawati Region. *Int. J. Sci. Eng. Res.* **2014,** *5*, 892–95.

Pareek, O. P.; Purohit, A. K. Patch Budding in Khejri (*Prosopis cineraria*). *Indian J. Hort.* **2002,** *59* (1) 89–94.

Pokhriyal, T. C.; Bhandari, H. C. S.; Negi, D. S.; Chaukiyal, S. P.; Gupta, B. B. Identification of Some Fast Growing Leguminous Tree Species for Nitrogen Fixation Studies. *Indian For.* **1990,** *116*, 6, 504–507.

Puri, S.; Kumar, A.; Singh, S. Management and Establishment of *Prosopis cineraria* in Hot Deserts of India. In *Integrated Land Use Management for Tropical Agriculture*, Proceedings of Second International Symposium Queensland, Sept 15–25, 1992, 246.1.46.

Puri, S., Kumar, A.; Singh, S. Productivity of *Cicer arietinum* (Chickpea) Under a *Prosopis cineraria* Agroforestry System in the Arid Regions of India In Special Issue*: Acacia* and *Prosopis cineraria. J. Arid Environ.* **1994,** *27*, 1, 85–98.

Purohit, A. K.; Samadia, D. K.; Pareek, O. P. Studies on Growth Enhancement of *Khejri* (*Prosopis cineraria* (L.) Druce) Seedlings and In Situ Budded Plants. *Annals Arid Zone* **2003,** *42* (1), 69–73.

Ramani, V.; Kant, U.; Phenolics and Enzyme Involved in Phenol Metabolism of Gall and Normal Tissues of *Prosopis cineraria* (Linn.) In Vitro and In Vivo. *Proc. Indian Nat. Sci. Acad. No. 5 & 6.* **1989,** 417–20.

Ramani, V.; Kant, U.; Quereshi, M. A. Auxin Profile of Gall and Normal Tissues of *Prosopis cineraria* (Linn.) Druce Induced by *Lobopteromyia prosopidis* Mani, In Vitro and In Vivo Proceedings: Plant Sciences. **1989,** *99,* 235–9.

Rani, B.; Singh, U.; Sharma, R.; Gupta, A.; Dhawan, N. G.; Sharma, A. K.; Sharma, S.; Maheshwari, R. *Prosopus cineraria* (L) Druce: A Desert Tree to Brace Livelihood in Rajasthan. *Asian J Pharm. Res. Health Care* **2013,** *5* (2) 58–64.

Rathore, M. Nutrient Content of Important Fruit Trees from Arid Zone of Rajasthan. *J. Hortic. For.* **2009,** *1* (7), 103–108.

Reddy, Y. V. R.; Sudha, M. Economics of Agroforestry System: CRIDAs Experience. In *Agroforestry Systems in India: Research and Development*; Singh, R. P., Ahlwat, I. P. S., Gangasaran, Eds.; Indian Society of Agronomy: New Delhi, 1989; pp 179–186.

Samadia, D. K.; Purohit A. K.; Pareek, O. P. Genetic Diversity in Vegetable Type Khejri. *Indian J. Agro For.* **2002,** *4* (2), 132–134.

Saroj, P. L.; Raturi, G. B.; Vashishtha, B. B.; *Biodiversity in Khejri (Prosopis cineraria) and its Utilization in the Arid Zone of Rajasthan*, In Proceedings of Resource Conservation and Watershed Management-technology Options and Future Strategies; Dhyani, et al. Eds; CSWRI: Dehradun, 2002, pp 123–129.

Satyanarayana, Y. *Habitat and Plant Communities of Indian Desert.* In Proceedings of Symposium Problems of Arid Zone; Ministry of Education, Government of India: New Delhi, 1964, 59–68.

Saxena, S. K. Vegetation and its Succession in the Indian Desert. In *Desertification and its Control*, Rear, New Delhi, 1977, pp 176–192.

Saxena, S. K.. Herbage Growth under Khejri Canopy. In *Khejri in the Indian Desert*; CAZRI: ICAR, 1980; pp. 4–9, 25–27; 64–67.

Saxena, S. K. Khejri (*Prosopis cineraria*) and Bordi *(Zizyphus nummularia*) Multipurpose Plants of Arid and Semi Arid Regions of India. In *Agroforestry in Arid and Semi Arid Zones*; CAZRI: Jodhpur, 1984, pp 111–123.

Shankar, V. Distribution of Khejri in Western in Khejri (*Prosopis cineraria*). In *Thar Desert its Role in Agro forestry*; Mann, H. S., Saxena, S. K., Eds.; CAZRI Monograph No. 11: Jodhpur, 1980, pp 11–19.

Singh, G. Impact of Diseases and Insect Pests in Tropical Forests Palacobotanocal; Features of the Thar Desert. *Ann. Arid Zone* **1970,** *8* (2), 188–195.

Solanki, K. R.; Kackar, N. L.; Jindal, S. K. Propagation of *Prosopis cineraria* (L) MacBride, by Air Layering. *Current Sci.* **1984,** *53*, 1166–1167.

Solanki, K. R.; Kackar, N. L.; Jindal, S. K. Propagation of *Prosopis cineraria* (L) MacBride. *Indian For.* **1986,** *112* (3), 202–207.

Srivastava, J. P.; Hetherington, J. C. Khejri (*Prosopis cineraria)*: A Tree for the Arid and Semi Arid Zones of Rajasthan. *Int. Tree Crop J.* **1971,** *7* (1–2) 125–127.

Vishal, N.; Pareek, O. P.; Saroj, P. L.; Sharma, B. D. Biodiversity of Khejri in Arid Region of Rajasthan: I-Screening of Khejri for Culinary Value. *Indian J. Soil Conserv.* **2000,** *28* (1), 43–47.

Vishnu Mittre, B. *Problems and Prospects of Palacobotanocal Approach Towards the Investigation of the History of Rajasthan Desert.* In Proceedings of Workshop on the Problems of Desert in India (Memeo), Geographical Survey of India, Jaipur, 1985.

Young, A. Agro forestry for Soil Conservation. International Council for Research in Agroforestry, Nairobi. 1989.

CHAPTER 24

NONCONVENTIONAL LEGUMES FORAGE CROPS

DHIMAN MUKHERJEE[1*] and MD. HEDAYETULLAH[2]

[1]*AICRP on Wheat and Barley, Directorate of Research, Bidhan Chandra Krishi Viswavidyalaya, Kalyani 741235, West Bengal, India*

[2]*Assistant Professor & Scientist (Agronomy), AICRP on Chickpea, Directorate of Research, Bidhan Chandra Krishi Viswavidyalaya, Kalyani 741235, West Bengal, India*

[*]*Corresponding author. E-mail: dhiman_mukherjee@yahoo.co.in*

ABSTRACT

Since ancient times, cattle breeding and milk production have been the important professions in India, and this directly relate with the availability of various forage and fodder crops. Free grazing was practiced and it became a way of life for healthy growth of cattle. Presently, livestock production is primarily based on rangeland grazing. The grazing activity is mainly dependent on the availability of the grazing resources from pastures and other pasture lands, namely, forests, miscellaneous tree crops and groves, cultivable wastelands, and fallow land. Various underutilized leguminous crops such as *Centrosema pubescens*, *Clitoria ternatea*, *Pueraria phaseoloides*, *Macroptilium atropurpureum*, *Mucuna pruriens*, lablab bean, *Desmodium*, etc.

24.1 INTRODUCTION

Agriculture and animal husbandry in India are interwoven with the intricate fabric of the society in cultural, religious, and economical ways as mixed farming and livestock rearing forms an integral part of rural living. Efficient utilization of limited land resources and other agricultural inputs for obtaining the best from the harvest in the form of herbage per unit area and time is the primary objective of intensive forage production system (Mukherjee, 2015). Since ancient times, cattle breeding and milk production have been the important professions in India, and this directly relate with the availability of various forage and fodder crops. Free grazing was practiced and it became a way of life for healthy growth of cattle, etc. Presently, livestock production is primarily based on rangeland grazing. The grazing activity is mainly dependent on the availability of the grazing resources from pastures and other pasture lands, namely, forests, miscellaneous tree crops and groves, cultivable wastelands, and fallow land. Various underutilized leguminous crops such as *Centrosema pubescens*, *Clitoria ternatea*, *Pueraria phaseoloides*, *Macroptilium atropurpureum*, *Mucuna pruriens*, lablab bean, and *Desmodium* are nutritious fodder. In integrated dairy, animals are dependent on underutilized fodder crops (Mukherjee, 2012). However, farmers may require fodder at different times during the season. But, subjecting to different cutting regimes could have an impact on crop yields through effects on N fixation, transfer, and mineralization, as well as changes in competition for light, nutrients, etc. Legume species are important supplements for livestock and at times can completely replace purchased feeds if used properly. There are many legumes species commonly used for pasture or fodder in India and world too. In this chapter, few important nonconventional leguminous forage crops are discussed which are suitable under various land topography and convenient to use as feed for cattle, poultry feed, etc.

24.2 CENTROSEMA PUBESCENS

Family: Fabaceae
Common name: Centro or butterfly pea.

C. pubescens is native to Central and South America; grows well along roadsides, in waste places, on river banks, and on coconut plantations. This is a perennial herb that can reach a height of 45 cm with deep taproot up to 30 cm; this character helps this plant to treat as drought-tolerant crop. Approximately 120–270 kg/ha nitrogen is fixed by this crop from atmosphere annually. Stems grow and branch rapidly, producing a dense mass of branches and leaves on the soil. Stems do not become woody until about 18 months after planting. Leaves are trifoliate, with elliptical leaflets approximately 4 cm × 3.5 cm, dark green and glabrous above. Flowers are generally pale violet with darker violet veins, born in axillary racemes. Fruit is a flat, long, dark brown pod 7.5–15-cm long, containing up to 20 seeds. Seeds are spherical and dark brown when ripe. This is a short-day plant.

24.2.1 UTILIZATION

C. pubescens is widely used as forage, and a good source of protein to grazing cattle. This is grown as a cover crop because it naturally suppresses weeds and is very tolerant to drought. For human consumption, it is not suitable but provides benefits through soil fertility and animal health. The leaves can also be used as a cheap source of protein for broiler chickens. It is a good source of calcium and potassium for animals. *C. pubescens* is a good source of protein (23.24% of crude protein in leaf), calcium (1.24% in leaf), and potassium for cattle as forage (Nworgu and Egbunike, 2013). It can be used to feed broiler chickens and broiler finishers as leaf meal in a quantity up to 20 g per day. More than that causes reduction in growth performance. This is a very cheap alternative to other sources of protein that are usually more expensive, such as, soybean (Nworgu and Egbunike, 2013).

24.2.2 CULTIVATION PRACTICES

This acts as a promising forage and as an alternative to enhance the protein content of livestock feed. This plant is well adapted to tropical conditions and altitudes below 600 m from sea level. Annual rainfall 1000–1800 mm is assumed good for its optimum production. Centro is unable to tolerate cold temperatures, but has very low water and rainfall requirements. It is easy to manage and improves soil nitrogen levels. This is propagated by seed, planted directly into the ground or broadcast over a field typically before the rainy season. Centro grows well in soils without fertilizer, since it is very adaptable to its environment. For optimal yields, it is best to grow in wet and humid sandy to clay soils with pH 4.9–5.5. Nodulation and nitrogen fixation are highly correlated with soil nature. It performs better on acidic soils than alkaline soils. Due to deep root system, it can take up water from a significant depth. It grows well in nutrient-poor soils.

However, fertile soils containing calcium, phosphorus, molybdenum, potassium, and copper increase yield to significant levels. This plant is also able to endure soils with a high level of manganese, and very capable of enduring waterlogging, flooding, and shade. For good yield, seed rate is 4–6 kg per hectare drilled in prior to the rainy season. Apply 1–2 kg of seed/ha when intercropped with grasses. For green manure, it can be sown up to 8 kg/ha. For broadcasting, increase the seeding rate. Seeds of *C. pubescens* have a mechanical dormancy that has to be broken by soaking the seeds for 3–5 min in water at 85°C. Other methods include chipping a small part of the seed case near the embryo, being very careful not to damage the seed, or, on a commercial scale, a short immersion in sulfuric acid. After the seeds have passed the dormancy-breaking treatment, they can be inoculated with *Rhizobium* and planted with a no-till planter. A typical seed planting depth is 2.5–5 cm. The shallower depth is used when the soil moisture is appropriate, but when the soil is dry the seed should be planted deeper to reach moisture. The plant has fairly high calcium content in the leaves, so addition of calcium can be important. Liming is good for this plant. Liming helps to enhance soil pH and supply calcium to the plant that will increase the calcium content in the leaves. *C. pubescens* does not need nitrogen fertilizer because it is supplied through the nodules. This is a tropical forage, so it requires very low phosphorus, but it responds to P fertilization. Leaves should be a minimum of 0.16% phosphorus at flower formation. The ideal available P in the soil for a good yield is between 2 and 5 mg of phosphorus per kilogram of soil and 12.4 mg per kilogram of soil of potash. *C. pubescens* can produce yield within range of 3.5–4.5 t per hectare of dry matter (DM). *C. pubescens* can be intercropped with various grasses. It grows well with *Panicum maximum*, *Hyparrhenia rufa*, *Melinis minutiflora*, *Chloris gayana*, *Pennisetum purpureum*, and *Paspalum dilatatum*; less successfully with *Brachiaria mutica*, and *Digitaria decumbens*. Sometimes it is planted with *Calopogonium mucunoides* and *P. phaseoloides* to give a quick cover. Since this plant is vigorously twining, it naturally suppresses weeds by creating a dense ground cover and is fairly good at spreading naturally to cover a large surface area. The combination of grass and centro is more suppressive of weeds than any other grass and legume combination. Insects are the biggest problem for this plant. Pests include meloidae beetles, thrips, red spiders, bean flies, and caterpillars. Centro is mostly unaffected by diseases, and tends not to have any major attacks.

24.2.3 HARVEST AND STORAGE

Seed harvest is usually performed by hand. Mechanical harvesting is difficult due to the plant architecture. When the plants are ripening, they are collected and spread to dry in the sun until they are ready to be threshed. After the seeds are removed from their pods, they are typically cleaned in hot water or with a chemical to eliminate any pathogens that may be present. Storage of this seed should be dry and free from humidity because wet environments give rise to pests and pathogens and promote their growth.

24.3 Pueraria phaseoloides

Family: Fabaceae
Subfamily: Faboideae

It is known as puero in Australia and tropical kudzu in most tropical regions. This crop is indigenous to east or Southeast Asia. Preferentially it is grown in plantation of cocoa or banana, at low altitudes (often under 600 masl) in wet evergreen or monsoon forests. *P. phaseoloides* is a deep-rooting perennial herb, building a subtuber, which allows to resist waterlogged soils and short periods of drought. The aboveground structure can grow up to 30 cm at day and often the stems can reach 20 m of elongation and climb over other plants or anthropogenic objects. The leaves are large trifoliate typical for Leguminosae. The growing season go from early spring to late fall in the subtropics and the year round in the tropics. Flowers color range from mauve to purple and the dimensions are small and disposed in scattered pairs on a raceme. Mature pods show a black color and hair coat. Each pod contains 10–20 seeds. This has high protein content in the seeds (12–20%).

24.3.1 UTILIZATION

P. phaseoloides is a promising forage and cover crop used in the tropics. The nutrient, protein (3.8%), and sugar (7.3%) content of the whole fresh plant (green part) is very high. Due to its rich nutrient content it has a good feeding value. About 100 g of *P. phaseoloides* contain 1880 kJ of energy, of which a big share is available as metabolic energy,

24.3.2 CULTIVATION PRACTICES

P. phaseoloides is capable to grow in a large soil spectrum. Acid soils are not a problem, and the pH tolerance is between 4.3 and 8. This is used as cover crop or as part of a mixture in pastures, its production methods differ for both uses. This forage crop is propagated by drill sowing mainly, where the distance between the drill rows is set to 1 m. Furthermore, it can be hand planted or propagated by cutting. For increased germination and sanitary protection, a hot water treatment (50–70°C) can be applied. For soil cover or green manure use, normally sown with a sowing density of 4.0 kg seed per hectare, which is similar to 32–35 seeds per m^2, depending of the seed weight. When sowing pastures with a high weed pressure, the number of seed can reach up to 70 seeds per m^2, when there is a high weed pressure. When used in mixture, this crop is sown with a density of 1.5–2.0 kg per hectare, which is equal to 12–18 seeds per m^2. Well root development is attained under presence of P, Mg, and Ca. Furthermore, they found the highest yield reduction under low-phosphate conditions followed by low Ca and Mg conditions, whereas at low K, N, and Na conditions yield was 50% lower. As a legume, *P. phaseoloides* can compensate for low N conditions by increasing symbiotic nitrogen fixation. This also explains its well response to added P. On poor soils, 100 kg of P_2O_5 showed to bring beneficial effect on the yield. The inoculation of the seeds before sowing with *Bradyrhizobium* is advised for primary cultivated areas. The management of *P. phaseoloides* grown in mixtures is challenging. This crop often shows a high palatability compared to tropical grasses and hence under high grazing pressure it can disappear. If the grazing pressure is too low, it can become dominating due to its fast growth and its climbing ability. Its growth is also affected by the other species in the mixture. It grows well with guinea and Napier grass. However, it cannot persist when grown with *Brachiaria decumbens* or pangola grass.

24.3.3 HARVEST AND STORAGE

When used as a forage crop, *P. phaseoloides* is mainly grazed. Cutting for hay, silage, and barn betting is possible as well. When used as a green manure, kudzu is directly incorporated into the soil. Harvesting of the seeds can be done by hand or with harvesting machines. When grown as monoculture, the yield can reach up to 10 t DM per hectare, whereas the biggest proportion of the yield is produced during wet season. The yield of mixtures can reach up to 23 t per hectare when grown under optimal conditions. The optimum seed yield can only be attained when harvested by hand. If the seeds are harvested with a machine, the harvested yield is noticeably lower.

24.4 Clitoria ternatea

Family: Leguminosae
Common names: Blue pea, cordofan pea, honte (French), and cunha (Brazil).

C. ternatea is widespread throughout humid and subhumid lowlands of Asia, the Caribbean, and Central and South America. This is forage of a consistently high nutritional value. This crop is moderately tolerant of salinity and sodicity. It may have wider application to smallholder farm systems in Southeast Asia. Its drought tolerance and adaptation to heavy clay soils, and the palatability and quality of its forage, suggest it could be used to improve natural grassland in extensive farm systems in the subhumid to semiarid tropics, given appropriate grazing management. Its susceptibility to close grazing or cutting is a major limitation to general use. This is a vigorous, strongly persistent, herbaceous perennial legume; stems are fine twining, sparsely pubescent, suberect at base, 0.5–3-m long. Leaves pinnate with 5 or 7 leaflets, petioles 1.5–3-cm long, stipules persistent, narrowly triangular, 1–6-mm long, sibilate, prominently three-nerved, rachis 1–7-cm long, stipels filiform. Flowers are axillary, single or paired, color ranges from white, mauve, light blue to dark blue. Chemical compounds isolated from *C. ternatea* include various triterpenoids, flavonol glycosides, anthocyanins, and steroids (Staples, 1-). It is propagated by seed and readily self-propagates and spreads under favorable conditions by seed thrown vigorously from the dehiscing dry pods. Seed is also spread in cattle dung. In traditional Ayurvedic medicine, it is ascribed various qualities including memory enhancing, nootropic, antistress, anxiolytic, antidepressant, anticonvulsant, tranquilizing, and sedative properties. In traditional Chinese medicine, and consistent with the Western concept of the doctrine of signatures, the plant has been ascribed properties affecting female libido due to its similar appearance to the female reproductive organ (Staples, 1992).

24.4.1 UTILIZATION

Originally used as a cover crop. Used for short and medium-term pastures and as green manure, cover crop, and protein bank. Increases soil fertility to improve yields of subsequent crops (maize, sorghum, and wheat), when grown as green manure or ley pasture. Also used for cut-and-carry and conserved as hay. This has excellent nutritive value with high protein and digestibility (up to 80%) with nitrogen concentrations of 3.0% N for leaf and 1.5% N for whole plant (Jones et al., 2000). Its crude protein content is comparable to that of alfalfa with values for the fresh forage typically higher than 18%DM. Forage quality persists even when maturity is advanced, without affecting digestibility or feed intake. For instance, protein content in *C. ternatea* hay varied from 23% DM in the vegetative state (42 days regrowth) to 19% DM at seeding (82 days). Unlike other legumes, this seems to be relatively free of toxic compounds and can be fed to ruminants and monogastric species, though its relatively high fiber content may be limiting for pigs and poultry. Very palatable, thus requires grazing management to persist (Cook et al., 2005).

24.4.2 CULTIVATION PRACTICES

Best results are achieved by planting in narrow rows (15–50 cm apart) at about 2–4 kg/ha seed rate for long-term pastures and about 6 kg/ha for short-term pastures to achieve plant densities of 5–10 plants/m^2. Excellent results can be achieved when sown as a crop using conventional planters and press wheels to achieve good soil/seed contact. For optimum yield as a green manure crop, use a seeding rate of 12 kg/ha. As a component of grass–legume pastures, can also be planted behind a blade plow. *C. ternatea* can be sown in pure stands (generally as a short-term rotation with crops) or in association with tall and tussock grasses for permanent pasture. When sown for pasture, it does well with elephant grass (*P. purpureum*), forage sorghums (*Sorghum bicolor*), millet species, Guinea grass (*Megathyrsus maximus*), pangola grass (*Digitaria eriantha*), gamba grass (*Andropogon gayanus*), or *Dichanthium aristatum*. When it is oversown in permanent pasture or sown in mixture with fast growing grasses, its establishment may be more difficult (Cook et al., 2005; Staples, 1992). In places where it is intended for revegetation, it can be sown with buffel grass (*Cenchrus ciliaris*) and Rhodes grass (*C. gayana*) (Cook et al., 2005).

Once established, this crop quickly covers the soil and can be directly harvested by grazing or as cut-and-carry forage. It should not be cut too low and too often. It is sensitive to trampling that may hamper regrowth from the tips, and cattle should not enter the stand more than 2–3 h/day. Only light grazing should be allowed during the establishment year so that the plants can set seed for stand regeneration and develop a strong frame that can withstand grazing. Heavy grazing is then possible, provided it is done rotationally (Cook et al., 2005). Fertilizer requirement is low, when sown on suitable soils, but P and S may be required on infertile soils. Weed competition will delay establishment but, once established, *Clitoria* can smother most weeds. Seed should be inoculated with *Rhizobium*. Mechanically scarify seed with a high hard seed content (>30%) when soil conditions favor immediate germination. Butterfly pea establishment is considered a much lower risk on heavy textured soils because of the large seed size and greater weed tolerance than alternatives such as leucaena (*Leucaena leucocephala*). Use of preemergent herbicide such as imazethapyr, 2–8 weeks prior to sowing is desirable to achieve successful control of weeds during establishment in old cropping areas.

24.4.3 HARVESTING AND YIELD

Hand harvest where economical, but can achieve 700 kg/ha by mechanical harvesting methods (direct heading). Irregular pod maturity affects best time of harvest as some pods will have shattered while flowers and green pods are still present. Forage DM may range from 0.2 to 16 t/ha/year depending on growing conditions. In dry Australian conditions, the cultivar Milgarra yielded 2–6 t DM/ha/year. Under irrigation, yields up to 30 t DM/ha could be achieved (Cook et al., 2005; Staples, 1992). Live weight gains of 0.7–1.3 kg/ha/day recorded for steers grazing pure *Clitoria* pastures in central Queensland, Australia.

24.5 Desmodium

Family: Fabaceae
Subfamily: Faboideae

Desmodium is a trailing or climbing perennial legume with small leaves and deep roots. In favorable conditions it forms dense ground cover. Most common of numerous varieties are greenleaf and silver leaf. Greenleaf desmodium [*Desmodium intortum* (Mill.) Urb.] is leafier, with reddish brown to purplish spots on the upper surface of the leaves and reddish brown stems. Silver leaf desmodium [*Desmodium uncinatum* (Mill.) Urb] has stems and leaves covered in dense hairs, which make them stick to hands and clothing. It has green and white leaves, light green underneath. The climatic range is similar for both species, but silver leaf desmodium is more frost tolerant than greenleaf desmodium because it flowers about a month earlier. Silver leaf is more tolerant to acid soils but less tolerant to drought than greenleaf. *Desmodium* is popular in cut-and-carry feeding systems. It can be intercropped with fodder crops such as maize or Napier grass and can help control weeds. Greenleaf desmodium is a large perennial tropical forage legume. It is a branched decumbent plant with long trailing and climbing pubescent stems that root at the nodes. Being an N-fixing legume, greenleaf desmodium can improve soil fertility. It can be used as ground cover as it needs only 4 months to cover the soil, and to prevent weeds from developing It has been used as ground cover on coffee plantations. The stems are green or sometimes red, 1.5–7.5-m long and about 7 mm in diameter. Greenleaf desmodium has many trifoliate leaves. The leaflets are ovate, 2–7-cm long × 1.5–5.5-cm broad, reddish-brown to purple in color. The flowers are borne on terminal compact racemes, deep lilac to deep pink in color. The pods are narrow, segmented, 5-cm long, and contain 8–12 kidney-shaped seeds that adhere strongly to hair or clothing. The seeds are about 3-mm long × 1.5-mm wide. Greenleaf desmodium is mainly used as a fodder legume. It can be grazed as a long-term pasture, cut and offered fresh in cut-and-carry systems, or cut from irrigated pastures for conservation as hay or silage. It is a valuable ground cover providing abundant leaf material that decomposes slowly in the soil (Cook et al., 2005). Greenleaf desmodium originated from Central and northwestern South America and is now widespread throughout the tropics. It became naturalized in small areas of the higher rainfall subtropics and elevated tropics. Greenleaf desmodium is a summer growing perennial.

24.5.1 UTILIZATION

The legume is very nutritious with high protein content and very palatable. It is resistant to grazing but special management is needed to regenerate before animals are brought back after grazing. If this is not done, the crop can easily disappear. *Desmodium* fodder contains 27% DM and 20.9% crude protein, it is as good as alfalfa as a protein supplement in poultry feedings.

24.5.2 CULTIVATION PRACTICES

Desmodium does well in warm, wet regions at altitudes of 800–2500 m that receive at least 875 mm of rainfall per year. It can be grown in areas where annual rainfall is above 900 mm and up to 3000 mm. Optimal temperature ranges between 25°C and 30°C. It grows well on slopes. During the growing season, it is more susceptible to drought and has better tolerance of flooding and waterlogging. *Desmodium* is a shade-tolerant forage legume, coming well under the shade of trees such as tamarind, coconut, and eucalyptus. Greenleaf desmodium can grow on a wide range of soils, provided they are not too acidic (pH above 4.5–5) and not saline. It is not tolerant of heavy frosts or fire (Kifuko-Koech et al., 2012). It is tolerant of shade and can be grown in coffee plantations. It is adapted to a wide range of soils from sands to clay loams and tolerates slight acidity but not salinity. Acidic soils can be improved by applying manure at the rate of 8 t/ha before sowing or planting. *Desmodium* can be grown as a pure stand or as a mixture with Napier grass in cut-and-carry plots. It can be grown under a maize crop or even as a cover crop under banana or coffee. It can be established by either seeds or cuttings. *Desmodium* seed is relatively expensive and very small. The seedlings can be swamped by weeds, so it is best sown in a weed-free, well-prepared nursery seedbed with fine-textured soil. A seedbed 3 m × 3 m, raised 15 cm, requires about 100 g of seeds. Greenleaf desmodium has very small seeds and requires a well prepared seedbed for establishment. It can be sown from spring to midsummer or later in frost-free areas. The best time to plant is at the start of rains. It is possible to propagate greenleaf desmodium by rooted cuttings. Once established, it grows vigorously and spreads rapidly into ungrazed areas because of its stolons. It grows well with a wide range of tussock grasses such as *Setaria* spp., *Megathyrsus maximus*, *P. purpureum*, *M. minutiflora*, *Pennisetum clandestinum*, or *Digitaria eriantha*. Greenleaf desmodium is a N-fixing legume that has been reported to fix 213–300 kg N/ha/year in the soil, but it transfers only 5% of this nitrogen to its companion grasses (Skerman et al., 1990). Greenleaf desmodium combines well with other legumes such as siratro (*M. atropurpureum*) or perennial soybean (*Neonotonia wightii*) (Cook et al., 2005). For areas with two rainy seasons, sow seeds during the short rains but plant cuttings during the long rains. Sow the seed immediately after adding the inoculant, either by drilling or by broadcasting. *Desmodium* can be grown between rows of Napier grass. When this crop is grown with Napier, the nitrogen it adds to the soil benefits the Napier and reduces the amount of nitrogen fertilizer required for topdressing. Greenleaf desmodium is very palatable and tends to be heavily grazed, but it cannot stand constant heavy grazing or frequent defoliation that removes the bud-promoting sites (cutting heights under 7–15 cm are recommended). After grazing, there should be enough vines and leaves left to allow good regrowth. Under careful grazing management, greenleaf desmodium pastures rarely persist for more than 6 years (Cook et al., 2005). Being a late flowering species, this provides good stand over feed during autumn and winter in frost-free areas (Cook et al., 2005). It can be cut for hay and makes good quality silage when mixed with molasses (8% fresh matter basis) (Skerman et al., 1990). Greenleaf desmodium has a relatively poor nutritive value for a legume. Its protein content is in the 12–21% DM range, with an average of 15.5% DM, and very high levels of fiber, including lignin (about 9% DM). Greenleaf desmodium has moderate needs for added fertilizers, only P, S, K, and Mo being required (Cook et al., 2005). Apply 500 g of phosphate fertilizer to the 3 $m^2 \times 3$ m^2 plot before sowing and mix thoroughly with soil. Alternatively, add 15 kg dry farmyard manure to the seedbed before planting. When growing desmodium with Napier grass, add one handful of farmyard manure per hole at planting and mix. Keep the plot weed free, especially during the early stages of establishment. Once *Desmodium* is fully established, it forms a complete groundcover that smothers the weeds, thus reducing the labor and cost of weeding the Napier plot.

24.5.3 HARVESTING AND YIELD

The crop must be harvested as soon as it starts flowering. If it is allowed for some more days after flowering; the stem gets hard and turns woody. Moreover, the leaves start falling. *Desmodium* pure stand starts harvesting after at least 4 months. The best harvesting schedule is to cut at 12-week intervals, cutting 10 cm or higher above soil level. DM yields of greenleaf desmodium range from 12 to 19 t/ha/year, which is higher than silver leaf desmodium (7–9 t DM/ha/year). Better yields are obtained with longer cutting intervals of 30–85 days (Ecocrop, 2014).

24.6 Macroptilium atropurpureum

Family: *Fabaceae* (alt. *Leguminosae*)
Subfamily: *Faboideae*

This has different names in different parts of world, such as atro, siratro, purple bean, purple bush bean (English), purpur bohne (German), and conchito (Spanish). Widely cultivated and naturalized throughout tropics and subtropics. Siratro is native to North America (USA, Texas) and Mexico. This is perennial herb with deep, swollen taproot and trailing, climbing, and twining stems. Stems at the base of older plants fibrous (>5 mm diameter), younger stems mostly 1–2 mm diameter, pubescent to densely pilose with white hairs, occasionally forming nodal roots under ideal conditions. Leaves are trifoliolate, leaflet blades 2–7 cm × 1.5–5 cm, darker green, and finely hairy on the upper surface. Inflorescence is a raceme comprising 6–12 often paired flowers on a short rachis, peduncle 10–30-cm long. Flower is 15–17-mm long, deep purple with reddish tinge near base. Pods are 5–10-cm long with 3–5-mm diameter, containing up to 12–15 seeds. Pods dehisce violently (shatter) when ripe. Seeds are speckled, light brown to black, and flattened ovoid (4 mm × 2.5 mm × 2 mm).

A rotational grazing system of 2 weeks on, 4 weeks rest has given good results in the humid subtropics. Stands deteriorate within a year or two under regular heavy grazing due to loss of bud sites, shorter plant life, reduced seed set, and declining soil seed reserves. Even with judicious management, they tend to become less productive over 4–6 years. While individual plants may persist for 4 years or more, pastures benefit from periodic spelling during seed set to bolster soil seed reserves and reinvigorate plants. This plant, well associate with erect or tussock grasses, and legumes require similar grazing management. It is less compatible with aggressive creeping grasses such as *Paspalum notatum* and *Digitaria eriantha*. Its companion species are, grasses such as *Chloris gayana*, *Cenchrus ciliaris*, *Panicum maximum*, *Setaria sphacelata*, with legumes mainly *Chamaecrista rotundifolia*, *Desmodium intortum*, *Macrotyloma axillare*, *Neonotonia wightii*, *Stylosanthes guianensis*, *S. hamata*, *Stylosanthes scabra*, and *Stylosanthes seabrana*. This crop is readily accepted, although cattle prefer fresh young grass early in the growing season. Siratro is heavily browsed by deer, and quail are attracted to the seed crop. Positive point with this crop is to be wide range of soil adaptation, drought resistant, high nutritive value, and palatability with good N fixation capability. However, it needs moderate fertility or added fertilizer. Intolerant of poor drainage, declines under grazing, and susceptible to leaf disease.

24.6.1 UTILIZATION

Mainly used for permanent and short-term pastures. While best suited to grazing, it can also be used for cut-and-carry or conserved as hay, is also used for soil conservation and as a cover crop, fallow crop (including after lowland rice), or as a forage crop sown with upland rice.

24.6.2 CULTIVATION PRACTICES

Occurs on a wide range of soils ranging from dark cracking clays, to yellow and red clays, to littoral dunes, red sands, and gravels with soil pH 5.2–8.5. It thrives in friable soils, but declines fairly rapidly in hard-setting soils. Tolerant of moderate levels of soil Al and Mn, and better tolerance of salinity than most tropical forage legumes. It is well adapted with rainfall between 700 and 1500 mm. Leaf disease can be a problem in higher rainfall environments. It is well adapted to drought, possessing a deep taproot and the ability to minimize evapotranspiration by virtue of pubescent leaves, and reduction in leaf size and shedding of leaves in response to the onset of dry conditions. It is intolerant of flooding or waterlogging. This is temperature sensitive crop. Good crop yield is obtained with day and night temperatures of 27–30°C and 24–25°C, respectively. Plant growth is poor below 18°C day and 13°C night (Jones and Jones, 2003). Leaves are burnt by light frost. More severe frosts kill the plant back to the crowns, but plants recover with the onset of warm, moist conditions. *M. atropurpureum* is fairly intolerant of shade and is best grown in full sunlight. Siratro seedlings do not establish under the dense shade of a closed canopy. Chemical composition and DM digestibility are largely unaffected by shade down to 24% sunlight (Jones and Jones, 2003). Seed scarification is most important for improvement of seed germination. Mechanical scarification can increase germination from 10% to 80%, treatment with concentrated sulfuric acid for 25 min for scarification. Seed is best sown at 1–2 kg/ha into a well-prepared seedbed. Seedlings are vigorous and can also establish with minimal cultivation. *M. atropurpureum* has a broad symbiont range for nodulation, and can nodulate effectively on native *Rhizobia* in most soils.

As per nutrient management concern, it prefers to grow under low nitrogen supply 10 kg/ha, phosphorus 20–30 kg/ha. Phosphorus plays an important role during establishment of plant, and for maintenance of 10–20 kg/ha/year, P is sprayed on soils (Jones et al., 2000). The need for other nutrients, particularly K and S, should be monitored by soil analysis or foliar deficiency symptoms. Micronutrient supply enhance its yield, particularly Mo 100–200 g/ha during establishment on deficient soils, and 100 g/ha Mo every 4–5 years as maintenance. Crude protein, mineral content, and digestibility can be increased by providing optimal P, S, and Mo.

24.6.3 HARVESTING AND YIELD

Seed can be either hand or machine harvested. For hand harvesting, ripe pods should be picked early in the day because as the day progresses, ripe pods dry out and dehisce violently with little stimulation. For larger scale commercial production, growth flushes are produced through irrigation with flowering occurring as moisture declines. Crops are then fairly synchronous, and can be direct headed when the majority of pods are ready to shatter. Seed yields vary greatly from 100–300 kg/ha. DM yields are mostly in the range of 5–10 t/ha/year, although yields are lower under more regular defoliation or grazing.

24.7 Mucuna pruriens

Family: Fabaceae
Subfamily: Faboideae

Common names of this plant are velvet bean, Mauritius velvet bean, and Bengal bean.

Velvet bean (*Mucuna* sp.) has become one of the key groups of species promoted for use as a legume cover crop, weed control, and green manure crop (Buckles, 1995). This is one of the most effective rotational crops for reducing nematode problems in cotton (*Gossypium hirsutum* L.), peanut (*Arachis hypogea*), and soybean (*Glycine max* L.). This grows as a leguminous vine and annual or sometimes short-lived perennial. This has vigorous, trailing or climbing, up to 6–18-m long. It has a taproot with numerous, 7–10-m long, lateral roots. The stems are slender and slightly pubescent. The leaves are generally slightly pubescent, alternate, trifoliolate with rhomboid ovate, 5–15-cm long × 3–12-cm broad, leaflets. The inflorescence is a drooping axillary raceme that bears many white to dark purple flowers. After flower pollination, velvet bean produces clusters of 10–14 pods. The velvet bean seeds are variable in color, ranging from glossy black to white or brownish with black mottling. Seeds are oblong ellipsoid, 1.2–1.5-cm long, 1-cm broad, and 0.5-cm thick (US Forest Service, 2011). This crop exhibits reasonable tolerance to a number of abiotic stress factors, including drought, poor soil fertility, and more soil acidity. Although this crop is sensitive to frost, growth occurs poorly in cold andwet soils (Burle et al.., 1992).

24.7.1 UTILIZATION

Velvet beans have three main uses: food, feed (forage and seeds), and environmental services. The plant can be a cover crop, and provides fodder and green manure. Velvet bean is a valuable fodder and feed legume. Vines and foliage can be used as pasture, hay, or silage for ruminants while pods and seeds can be ground into a meal and fed to both ruminants and monogastrics as a source of rich protein (Chikagwa-Malunga et al., 2009).

24.7.2 COVER CROP AND SOIL IMPROVER

Velvet bean is mainly grown as a cover crop and green manure because it can establish very quickly without requiring complete soil preparation (Cook et al., 2005). The main attributes of velvet bean are its fast growth and its long growing season in frost-free environments. It is thus possible for velvet bean to protect the soil through the wet monsoon season (Cook et al., 2005). Velvet bean is an N-fixing legume that has no specific *Rhizobium* requirements, but N fixation is favored by warm temperatures. As leguminous species, velvet bean is reported to improve soil fertility: it provides more than 10 t DM aboveground biomass/ha, and below ground it fixes some 331 kg N/ha equivalent to 1615 kg ammonium sulfate/ha (Cook et al., 2005; Wulijarni-Soetjipto et al., 1997).

24.7.3 NUTRITIONAL ATTRIBUTES

Velvet been forage contains 15–20% protein (DM basis) (Sidibé-Anago et al., 2009). Seeds are rich in protein (24–30%), starch (28%), and gross energy (10–11 MJ/kg) (Pugalenthi et al., 2005). They also contain desirable amino acids, fatty acids and have a good mineral composition. This plant has certain antinutritional factors such as L-dopa (from this dopamine, a potent anti-Parkinson disease agent), number of alkaloids notably mucunain, prurienine, and serotonin, trypsin and chymotrypsin (inhibiting activities, decrease protein digestibility, and induce pancreatic hypertrophy and hyperplasia), etc. (Buckles, 1995). Many treatments have been proposed to decrease the content in antinutritional factors of the seeds, such as, boiling in water for 1 h, autoclaving for 20 min, water-soaking for 48 h and then boiling for 30 min, or soaking the cracked seeds for 24 h in 4% $Ca(OH)_2$ (Cook et al., 2005; Pugalenthi et al., 2005).

24.7.4 CULTIVATION PRACTICES

Velvet bean originated from southern Asia and Malaysia and is now widely distributed in the tropics. This is found from sea level up to an altitude of 2100 m (Ecocrop, 2011). Velvet bean thrives best under warm, moist conditions, and in areas with plentiful rainfall. In such environments, its vines can grow up to 30 ft (Burle et al., 1992). However, specific growth characteristics depend on the genotype. It requires a hot moist climate with annual rainfall ranging from 650 to 2500 mm and a long frost-free growing season during the wet months. It can grow on a wide range of soils, from sands to clays but thrives on well-drained, light textured soils of appreciable acidity. Due to its large seeds, the crop does not require a lot of land preparation. Application of 500–700 kg/ha lime (preferably dolomitic lime on sandy soils) is recommended to encourage nodulation and efficient use of fertilizers. For soils with very low nitrogen level, a single application of 24 kg N/ha will be necessary to boost plant growth, 3–4 weeks after germination. This will thrive on soils where available soil phosphorus is low. Application of 200–250 kg/ha single superphosphate is sufficient for optimum herbage and seed production. Farmers are advised to keep the crop weed free by weeding as soon as weeds appear. This will also reduce pest infestation.

24.7.5 PLANTING

Seed is sown at a rate of 35–24 kg/ha in single crops at the beginning of the wet season, using interrow spacing of 0.9–1 m and within row spacing of 30–24 cm. A lower seed rate (wider spacing) is advisable in semiarid conditions, to reduce competition for moisture. Mucuna seeds are large and should be planted at a depth of 3–7 cm.

Intercropping with maize or sorghum: the velvet bean is a very vigorous climber. Therefore, it should be planted in-between cereals 3–4 weeks after they emerge (depending on predicted annual rainfall), ideally after the first hand weeding, if farmers are not using herbicides. If planted too early and densely, it can choke the cereal, thereby reducing cereal yield. Planting the velvet bean within the same row as maize and in-between the maize plants facilitates weeding and spraying. However, delaying the planting of legume for more than 4 weeks after sowing cereals may result in shading by the cereal crop and severe reduction in legume yield.

24.7.6 HARVESTING AND YIELDS

Harvesting of velvet bean pods can start as soon as they start turning from green to dark brown or black. When velvet bean is intended for forage, it may be harvested when the pods are still young, usually between 90–120 days after sowing (Wulijarni-Soetjipto et al., 1997). Harvesting at about 120 days after planting resulted in the best combination of biomass yield and nutritive value. Because of its dense-matted growth, velvet bean is difficult to harvest and cure for hay. Yields of hay range between 2.8 and 3.6 t/ha (Ecocrop, 2011). This crop is suitable in intercropping systems where it is grown with maize (Cook et al., 2005), pearl millet, sorghum, or sugarcane for support (Göhl, 1982). The crop gives reliable yields in dry farming and low soil fertility conditions that do not allow the profitable cultivation of most other food legumes. Velvet bean yields range from 10 to 35 t green material/ha and from 250 to 3300 kg seeds/ha depending on the cultivation conditions (Ecocrop, 2011).

24.8 LABLAB OR DOLICHOS BEAN (*Lablab purpureus*)

Family: Fabaceae

This is one of the most ancient crops among cultivated plants. This is also known as hyacinth bean or field bean. It is a bushy, semierect, perennial herb showing no tendency to climb. It is mainly cultivated either as a pure crop or mixed with finger millet, groundnut, castor, corn, bajra, or sorghum in Asia and Africa. It is a multipurpose crop grown for pulse, vegetable, and forage. The crop is grown for its green pods, while dry seeds are used in various vegetable food preparations. It is also grown in home gardens as annual crop or on fences as perennial crop. It is one of the major sources of protein in the diets in southern states of India. The consumer preference varies with pod size, shape, color, and aroma (pod fragrance). It is also grown as an ornamental plant, mostly in the United States for its beautiful dark green, purple-veined foliage with large spikes clustered with deep violet and white pea-like blossoms.

24.8.1 UTILIZATION

This crop is mainly used as fodder or green manure. As forage, it is very palatable, either as green fodder or as silage. It improves the soil condition with good ground cover. Young immature pods are cooked and eaten like green beans (older pods may need to be destringed). They have a strong flavor and some people like to mix them with other beans or green vegetables. Young leaves are eaten raw in salads and older leaves are cooked like spinach. Flowers are eaten raw or steamed. The large starchy root tubers can be boiled and baked. The immature seeds can be boiled and eaten like any shelly bean. Dried seeds should be boiled in two changes of water before eating since they contain toxins *Cyanogenic glucosides*. In Asia, the mature seeds are made into tofu and fermented for tempeh. They are also used as bean sprouts. Raw dry seeds are poisonous and can cause vomiting and even convulsions and unconsciousness.

24.8.2 CULTIVATION PRACTICES

Lablab is remarkably adaptable to wide areas under diverse climatic conditions such as arid, semiarid, subtropical, and humid regions where temperatures vary between 24°C and 35°C, low lands and uplands and many types of soils and the pH varying from 4.4 to 7.8. Loam, silty loam, and clay loam soils are best suited for Indian bean. Being a legume, it can fix atmospheric nitrogen to the extent of 170 kg/ha, besides leaving enough crop residues to enrich the soils with organic matter. It is a drought-tolerant crop and grows well in dry lands with limited rainfall. The crop prefers relatively cool seasons (temperature ranging from 14°C to 28°C) with the sowing done in July–August. It starts flowering in short days (11–11.5 h day length) and continues indeterminately in spring. Hyacinth bean flowers throughout the growing season.

Plow the land to a fine tilth with 5–6 plowing, form ridges and furrows 60 cm apart for bush types. Seed rate of 25 kg/ha for bush type and 5 kg/ha for pandal type is required. Treat the seeds with suitable *Rhizobial* culture, using rice gruel as binder. Dry the treated seeds in shade for 15–30 min before sowing. The seeds may be sown in rows or on ridges by drilling or by dibbling. Flat bed, ridges, and furrow layout is used with spacing of 90 cm × 90 cm. However in some places, in pandal system, dig pits of 30 cm × 30 cm × 30 cm at required spacing and fill them up with farmyard manure (FYM). Dibble single seed 30 cm apart on one side of the ridge formed at a spacing of 60 cm for bush type. For pandal type, sow 2–3 seeds/pit at 2 m × 3 m spacing. Hoeing and weeding can be done as and when necessary. Provide stakes to reach pandal of 2 m height and train the vines on pandal.

24.8.2.1 APPLICATION OF FERTILIZERS

The manures and fertilizers requirement under irrigated and dry conditions is given in Table 24.1.

24.8.2.1.1 Basal Dressing for Bush Type (Table 24.1)

TABLE 24.1 Basal Dressing for Bush Type.

Manures and fertilizers	Irrigated	Dry
FYM (t/ha)	12	10
N (kg/ha)	20	14
P_2O_5 (kg/ha)	24	30
K_2O (kg/ha)	20	20

FYM, farmyard manure.

24.8.2.1.2 For Pandal Type

Apply 10 kg FYM per pit (20 t/ha), 100 g of NPK 6:12:12 mixture as basal, and 10 g N per pit after 30 days. Apply 2 kg each of *Azospirillum* and *Phosphobacterium* per hectare at the time of sowing.

24.8.3 HARVESTING AND YIELD

The Indian bean becomes ready for harvesting after 2½–3 months of sowing. Full grown bean is harvested according to the need. The average yield is 100–120 q of green pods per hectare.

KEYWORDS

- **underutilized fodder crops**
- **agronomic management**
- **yield**
- **utilization**

REFERENCES

Buckles, D. Velvet Bean: A "New" Plant with a History. *Econ. Bot.* **1995,** *49*, 13–25.

Burle, M. L.; Suhet, J.; Pereira, D. V. S.; Resck, J. R. R.; Peres, M. S.; Cravo, W.; Bowen, D. R.; Bouldin, A.; Lathwell, J. D. Legume Green Manures: Their Dry Season Survival and the Effect on Succeeding Maize Crop. Soil Management Collaborative Research Support Program: Raleigh, North Carolina, *Soil Management CRSP Bulletin No*. 92-04, 1992.
Chikagwa-Malunga, S. K.; Adesogan, A. T.; Sollenberger, L. E.; Badinga, L. K.; Szabo, N. J.; Litell, R. C. Nutritional Characterization of *Mucuna pruriens*. In Vitro Ruminal Fluid Fermentability of *Mucuna pruriens*, Mucuna L-dopa and Soybean Meal Incubated with or Without L-dopa. *Anim. Feed Sci. Technol.* **2009,** *148*, 51–67.
Cook, B. G.; Pengelly, B. C.; Brown, S. D.; Donnelly, J. L.; Eagles, D. A.; Franco, M. A.; Hanson, J.; Mullen, B. F.; Partridge, I. J.; Peters, M.; Schultze-Kraft, R. *Tropical Forages*; CSIRO, DPI & F(Qld), CIAT and ILRI: Brisbane, Australia, 2005; pp 54–55.
Ecocrop. *Ecocrop Database*; FAO: Rome, Italy, 2014.
Jones, R. M.; Jones, R. J. Effect of Stocking Rates on Animal Gain, Pasture Yield and Composition, and Soil Properties from Setaria-nitrogen and Setaria-legume Pastures in Coastal South-east Queensland. *Trop. Grassl.* **2003,** *37*, 36–41.
Jones, R. M.; Bishop, H. G.; Clem, R. L.; Conway, M. J.; Cook, B. G.; Moore, K.; Pengelly, B. C. Measurements of Nutritive Value of a Range of Tropical Legumes and Their use in Legume Evaluation. *Trop. Grassl.* **2000,** *34*, 78–90.
Kifuko-Koech, M.; Pypers, P.; Okalebo, J. R.; Othieno, C. O.; Khan, Z. R.; Pickett, J. A.; Kipkoech, A. K.; Vanlauwa, B. The Impact of Desmodium spp. and Cutting Regimes on the Agronomic and Economic Performance of Desmodium–Maize Intercropping System in Western Kenya. *Field Crops Res*. 2012, *137* (1), 97–107.
Mukherjee, D. Studies on Profitability of Efficient Farming System in Midhills Situation of Eastern Himalaya. *J. Farming Sys. Res. Dev*. **2012,** *18* (1), 16–21.
Mukherjee, D. Food Security: A World Wide Challenge. *Res. Rev. J. Agric. Allied Sci.* (RRJAAS) **2015,** *4* (1), 3–5.
Nworgu, F. C.; Egbunike, L. Nutritional Potential of *Centrosema pubescens, Mimosa invisa* and *Pueraria phaseoloides* Leaf Meals on Growth Performance Responses of Broiler Chickens. *Am. J. Exp. Agric.* **2013,** *3* (3), 506–519.
Pugalenthi, M.; Vadivel, V.; Siddhuraju, P. Alternative Food/Feed Perspectives of an Underutilized Legume *Mucuna pruriens* var. utilis: A Review. *Plant Foods Hum. Nutr.* **2005,** *60*, 201–218.
Sidibé-Anago, A. G.; Ouedraogo, G. A.; Kanwé, A. B.; Ledin, I. Foliage Yield, Chemical Composition and Intake Characteristics of Three *Mucuna* Varieties. *Trop. Subtrop. Agroecosyst.* **2009,** *10*, 75–84.
Skerman, P. J.; Riveros, F. *Tropical Grasses*; FAO Plant Production and Protection Series No. 23; FAO: Rome, 1990.
Staples, I. P. *Clitoria ternatea* L. In *Plant Resources of South-east Asia No. 4.Forages*; Mannetje, L., Jones, R. M., Eds.; Pudoc Scientific Publishers: Wageningen, the Netherlands, 1992; pp 65–74. (pp 94–96.)
Wulijarni-Soetjipto, N.; Maligalig, R. F. *Mucuna pruriens (L.) DC. cv. Group Utilis. Record from Prosea base*; Faridah Hanum, I., van der Maesen, L. J. G., Eds.; PROSEA (Plant Resources of South-east Asia) Foundation: Bogor, Indonesia, 1997.

CHAPTER 25

AZOLLA: AN UNCONVENTIONAL FORAGE CROP

DULAL CHANDRA ROY[1*], ABHIJIT SAHA[2], and SONALI BISWAS[3]

[1]*Department of ILFC, WBUAFS, Mohanpur, Nadia 741252, West Bengal, India*

[2]*Department of Agronomy, College of Agriculture, Lembucherra 799210, Tripura, India*

[3]*Department of Agronomy, Bidhan Chandra Krishi Viswavidyalaya, Mohanpur, Nadia 741252, West Bengal, India*

**Corresponding author. E-mail: dcroy09@gmail.com*

ABSTRACT

Azolla is an aquatic fern which forms a symbiotic association with a blue-green algae *Anabaena azollae.* It is rich in carbohydrate, protein, minerals, vitamin, and essential amino acids. So it is also known as "green gold mine" of the nature. Previously it was used as biofertilizer in the paddy field as an alternative source of nitrogen and biomass. But nowadays, it is used in as a feed supplement for variety of animals such as cattle, goat, pigs, rabbits, chickens, ducks, and fish. It is also used in soil and water reclamation, biogas production, medicine preparation, and even as human food. In livestock sector a good amount of feed can be replaced with *Azolla* either in fresh or in dried powder form without hampering the production and quality of the product. Thus, it reduces the production cost and thereby increases the net profit of the farmers. *Azolla* can be collected from its natural habitat such as swamp, pond, and ditches or can be cultivated in concrete or semi-concrete pit or container, earthen pots, etc.

25.1 BOTANICAL CLASSIFICATION

Division: Pteridophyta
Class: Polypodiopsida
Order: Salviniales
Family: Salviniaceae
Genus: *Azolla*
Subgenus: (1) *Euazolla*
(2) *Rhizosperma*

The subgenus *Euazolla* consists of four species, namely, *Azolla filiculoides* (Lain.), *Azolla caroliniana* (WiUd.), *Azolla microphylla* (Kaulf.), *and Azolla mexicana* (Presl); while subgenus *Rhizosperma* has only two species, *Azolla pinnata* (R. Br.) and *Azolla nilotica* (Decne).

25.2 COMMON NAME

Ferny-*Azolla*, mosquito fern, water velvet.

25.3 BOTANICAL NAME

Azolla pinnata (R. Br.), *A. microphylla* (Kaulf.), *A. caroliniana* (WiUd.).

25.4 INTRODUCTION

Azolla is an unconventional forage crop widely used as a feed supplement either fresh or in dried powder forms for ruminants and nonruminants type of livestock. The term *Azolla* was first coined by Lamarck in 1783. The name *Azolla* is derived from the two Greek words, *Azo* (to dry) and *Ollyo* (to kill) thus it means that the fern is killed by drought.

Azolla is commonly used as biofertilizer as an alternative source of nitrogen and organic matter in the paddy field. *Azolla* is generally known as mosquito fern, duckweed fern, fairy moss, and water fern. *Azolla–Anabaena* is a symbiotic complex in which the endophytic blue-green algae *Anabaena zollae* lives within the leaf cavities of the water fern

Azolla (Lain). *Anabaena azollae* can fix sufficient amount of atmospheric nitrogen and supplies it to its host. The fern on the other hand provides a protected environment for the algae and also supplies it with a fixed carbon source.

25.5 ORIGIN AND GEOGRAPHICAL DISTRIBUTION

Place of origin of *Azolla* is in Vietnam and is believed to be domesticated as early as 11th century. *Azolla* is a small free-floating aquatic fern native to Asia, Africa, and the Americas. It is generally found in freshwater habitats in tropical, subtropical, and warm-temperate regions throughout the world. *A. pinnata* is found mostly in eastern and northeastern India but does not thrive well in hilly areas. *A. caroliniana* is a cold tolerant species and found in temperate regions of India. It can survive well even at very low temperature of 5°C during winter. *Azolla microphylla* is generally found in southern part of the country.

25.6 MORPHOLOGY

Leaves of most of the Indian species of *Azolla* are typically triangular in shape. *Azolla* macrophyte, called a frond, ranges from 1 to 2.5 cm in length in species such as *A. pinnata* and 15 cm or more in the largest species, *A. nilotica*. It consists of a main rhizome and it branches into secondary rhizomes, all of which bear small leaves alternately arranged. Unbranched adventitious roots hang down into the water from nodes on the ventral surfaces of the rhizomes. The roots absorb nutrients directly from the water, though in very shallow water they may touch the soil, deriving nutrients from it. Each leaf consists of two lobes: an aerial dorsal lobe, which is chlorophyllous, and a partially submerged ventral lobe, which is colorless and cup-shaped and provides buoyancy. Each dorsal lobe contains a leaf cavity which houses the symbiotic blue-green algae, the *Anabaena–Azollae* (Kannaiyan and Kumar, 2006). Interior surface of leaf cavity is covered by mucilaginous layer and is embedded with filaments of *Azolla* and permeated by multicellular transfer hairs.

25.7 ENVIRONMENTAL REQUIREMENTS

Azolla can be cultivated throughout year, though growing season and day length greatly affect its biomass production. Growing seasons of *Azolla*, however, are linked with other factors like pH, nutrient content, temperature, humidity, salinity, etc. Production of biomass during summer season is higher than other season of the year. *Azolla* generally prefers neutral to slightly acidic environment. Optimum pH level for growth of *Azolla* is 4.5–7.5, though it can survive in pH ranging from 3.5 to 10. Temperature is one of the most important growth regulating factors in *Azolla*. It can survive well in wide range of temperature variation. However, optimum temperature for rapid growth of biomass is 20–30°C. Very high temperature (above 35°C) and very low temperature (below −4°C) can inhibit the growth and production of biomass of *Azolla*.

Relative humidity is also an important factor for the growth of *Azolla*. Optimum humidity for better growth of *Azolla* is 70–75%. Relative humidity below 60% hampers the growth of *Azolla* and makes it dry and fragile. Among the nutrients, phosphorus plays vital role regarding growth of *Azolla* and production of green biomass. In various experiments, it has been found that increasing phosphorus (in the form of phosphate) supply and/or plant density resulted to increased sporulation. *Azolla* is sensitive to salinity to some extent. Some experiments have shown negative effect of salinity on growth of different species of *Azolla*. Masood et al. (2006) reported that salinity significantly inhibits growth and biomass production of *A. pinnata* and *A. filiculoides*. It also revealed that salt concentrations above 10 mM NaCl inhibited growth of *A. filiculoides*, while that of *A. pinnata* was stopped at 40 mM NaCl. Photosynthesis of *Azolla* is greatly affected by light intensity. In high light intensity and temperature it becomes red or brownish red. During hot summer months, a partial shade is to be given over the *Azolla* pit to protect it from the direct sunlight. In extreme cold, it also becomes red or brownish red in color and growth is also retarded.

25.8 NATURAL HABITAT

Azolla generally grows well in swamp, ponds, roadside ditches, lakes, and rivers where water is not turbulent.

25.9 CULTIVABLE SPECIES OF *Azolla*

In Indian agroclimatic situation, out of the six species of *Azolla*, only three species of *Azolla* can be cultivated, namely, *A. Pinnata*, *A. Microphylla*, and *A. Caroliniana*.

25.10 CULTIVATION OF *Azolla*

Azolla can be cultivated in a concrete or semi-concrete pit, plastic/fiber container even in earthed pot. A low-cost pit using bricks/soil, polythene, etc can also be made. Generally, a rectangular pit of size 2 m × 1 m with 25-cm depth is made. A polythene sheet is used to cover the pit and put weight on it using bricks or soil. Then 25 kg of well-sieved fertile soil is uniformly spread over it and 5 kg of 3–4 days old cow dung with 30 g of single superphosphate (SSP) fertilizer are added to the soil. Pour water into the pit and raise the water level about 10 cm from the soil level. Afterwards, fresh and pure *Azolla* is placed in the pit at the rate of 500 g/m^2. *Azolla* grows rapidly and covers the water surface within 8–10 days

25.11 INTERCULTURAL OPERATION

One-third of water of the pit is to be replaced with fresh water once in every 2 weeks. Also, 1 kg cow dung and 15 g SSP are to be added in the pit in every 2 weeks, after replacing the water. One-third of soil is to be replaced with fresh fertile soil once in every 2 months to maintain the nitrogen built up which enhances the growth of *Azolla*.

25.12 HARVESTING AND YIELD

Under ideal condition of growth, *Azolla* can be collected daily at the rate of 0.5–1.0 kg fresh biomass per m^2. After collection it should be washed with clean water twice or thrice to remove the off smell of cow dung and the washed water is poured back to the pit.

25.13 NUTRITIONAL QUALITIES

Azolla contains good amount of carbohydrate, protein, crude fiber, vitamin (vitamin A, vitamin B12, beta-carotene), minerals such as Ca, K, Fe, Mg, Zn, and almost 18 types of essential amino acids in sufficient quantity. *Azolla* is known as "green gold mine" of the nature for its high nutritive values. Nutrient content of *Azolla* is as follows (Tables 25.1–25.3).

TABLE 25.1 Proximate Analysis of *Azolla* (*Azolla* spp.), as Fresh.

Main analysis	Unit	Average value
Dry matter	% as fed	6.7
Crude protein	% DM	25.6
Crude fiber	% DM	15.0
NDF	% DM	43.8
ADF	% DM	31.8
Lignin	% DM	11.4
Ether extract	% DM	3.8
Ash	% DM	15.9
Starch (polarimetry)	% DM	4.1
Gross energy	MJ/kg DM	17.0

ADF, acid detergent fiber; DM, dry matter; NDF, neutral detergent fiber.

TABLE 25.2 Mineral Content of *Azolla*.

Minerals	Unit	Average value
Calcium	g/kg DM	11.0
Phosphorus	g/kg DM	6.1
Potassium	g/kg DM	17.4
Sodium	g/kg DM	9.0
Magnesium	g/kg DM	5.0
Manganese	mg/kg DM	762
Zinc	mg/kg DM	38
Copper	mg/kg DM	16
Iron	mg/kg DM	2500

DM: dry matter.

TABLE 25.3 Amino Acid Content of *Azolla* (*Azolla* sp.).

Amino acids	Unit	Average value
Alanine	% protein	6.4
Arginine	% protein	5.9
Aspartic acid	% protein	9.3
Cystine	% protein	1.6
Glutamic acid	% protein	12.6
Glycine	% protein	5.6
Histidine	% protein	2.1
Isoleucine	% protein	4.5
Leucine	% protein	8.4
Lysine	% protein	4.7
Methionine	% protein	1.4
Phenylalanine	% protein	5.4
Proline	% protein	4.9
Serine	% protein	4.5
Threonine	% protein	4.7
Tryptophan	% protein	1.8
Tyrosine	% protein	3.6
Valine	% protein	5.5

25.14 UTILIZATION

25.14.1 *Azolla AS BIOFERTILIZER IN RICE CULTIVATION*

Azolla is cultivated in the paddy field either as monocrop or as intercrop and incorporated into the mud/soil for increasing humus and nutrient content of the soil. This practice of *Azolla* cultivation is widely popular in the countries of Southeast Asia such as India, China, Philippines, Indonesia, etc. When *Azolla* used as a biofertilizer in paddy field it produces around 300 t of green bio-hectare per year under normal subtropical climate which is comparable to 800 kg of nitrogen (1800 kg of urea).

Azolla has quick decomposition rate in soil and thus it speedup the efficient availability of its nitrogen to rice plant. The quick multiplication rate and rapid decomposing capacity of *Azolla* has become paramount important factor to use as green manure cum biofertilizer in rice field. Basal application of *Azolla* at the rate of 10–12 t per hectare in the rice field increases soil nitrogen by 50–60 kg per hectare and reduces 30–35 kg of nitrogenous fertilizer requirement. Release of green *Azolla* twice as dual cropping in rice crop at the rate of 500 kg per hectare enriches soil nitrogen by 50 kg per hectare and reduces nitrogen requirement by 20–30 kg per hectare. Use of *Azolla* increases rice yield by 20–30%.

25.14.2 Azolla AS NUTRITIONAL SUPPLEMENT FOR LIVESTOCK

Azolla is used as feed supplement for variety of animals such as cattle, goat, pigs, rabbits, chickens, ducks, and fish. Seulthrope (1967) for the first time conducted an experiment and reported that *Azolla* can be utilized as fodder for cattle and pigs. From various experiments it was also found that broilers feeds supplemented with *Azolla* resulted better growth and body weight than the conventional feed (FAO, 2016). Murthy et al. (2013) fed milching with 2 kg fresh *Azolla* per day replacing 50% of concentrate during 3 months and reported the decrease in feed + labor costs by 16.5% and milk production costs by 18.5%, without hampering the milk production. Tamang et al. (1993) conducted an experiment on black Bengal goat by replacing its concentrate feed up to 50% with sun-dried *Azolla* and found that replacement with dried *Azolla* up to 20% maintained good health without any adverse effect. In case of cattle feed replacement to the tune of 30–40% with fresh or dried *Azolla* can give better milk yield and net profit.

Parthasarathy et al. (2002) reported that 5% feed replacement with dried *Azolla* gave better meat yield in Indian broiler chicken. Rai et al. (2012) conducted an experiment by grazing layer chicken with fresh *A. pinnata* and obtained higher body weight at 8 weeks or higher egg production at 40 and 72 days than control. In case of poultry feed replacement to the tune of 20–25% with fresh or dried *Azolla* can give better meat and egg production. In rabbit and pig, up to 50% level feed replacement with fresh or dried *Azolla* can give better profit without hampering the normal health of the animal.

Fresh *Azolla* is also a good nutrient source of freshwater fish cultivation. Digested *Azolla* slurry remaining after biogas production was suitable as fish pond fertilizer.

25.14.3 OTHER USES OF Azolla

Azolla can also be used in the control of mosquitoes. Thick *Azolla* mat on the water surface can prevent breeding and adult emergence from the stagnant water sources such as pools, ponds, wells, rice fields, and drains. Anaerobic fermentation of *Azolla* (or a mixture of *Azolla* and rice straw) produces methane gas which can be utilized as fuel and remaining effluent or slurry can be used as a fertilizer. *Azolla* has remarkable ability to remove heavy metals like Cu, Cd, Cr, Ni, Pb, etc. directly from pollutants or sewage water and thus it can be used to treat water polluted with heavy metals (Wagner, 1997).

Although, *Azolla* is relatively sensitive to salt, cultivation in saline environment for a period of two consecutive years decreased salt content from 0.35 to 0.15 and desalinate rate (71.4%) was 1.8 times faster than through water leaching and 2.1 times faster than *Sesbania* and also reduced the electrical conductivity, pH of acidic soil and increased calcium content of soil (Anjuli et al., 2004). *Azolla* has some medicinal uses also. In some African countries like Tanzenia, *Azolla* is used for preparing cough medicine. *Azolla* is also used for preparing various human foods such as *Azolla* soup, *Azolla* hardtack, *Azolla* ball, *Azolla* bread, etc. *Azolla* protein (20–30%) is close to that of soybean and is easily digestible.

Research by Katayama et al. (2008) in collaboration with Space Agriculture Task Force suggested *Azolla* as a component of the space diet during habitation on Mars and found that *Azolla* was found to meet human nutritional requirements on Mars.

FIGURE 25.1 **(See color insert.)** Various types of pits for *Azolla* cultivation.

KEYWORDS

- *Azolla*
- forage
- nutritional quality
- utilization
- biofertilizer

REFERENCES

Anjuli, P.; Prasanna R.; Singh P. K. Biological Significance of *Azolla* and Its Utilization in Agriculture. *Proc. Indian Natl. Sci. Acad.* **2004,** *70*, 299–333.

FAO. Feedipedia: An Online Encyclopedia of Animal Feed. www.feedipedia.org (accessed Aug 27, 2016).

Kannaiyan, S.; Kumar, K. Biodiversity of *Azolla* and Its Algal Symbiont, *Anabaena azollae*. In *NBA Scientific Bulletin Number-2*; National Biodiversity Authority: Chennai, Tamil Nadu, 2006; pp 1–31.

Katayama, N.; Masamichi Y.; Yoshiro K.; Chung C. L.; Watanabe I; Hidenori, W. Space Agriculture Task Force: *Azolla* as a Component of the Space Diet During Habitation on Mars. *Acta Astronaut.* **2008,** *63*, 1093–1099.

Masood, A.; Shah, N. A.; Zeeshan, M.; Abraham, G. Differential Response of Antioxidant Enzymes to Salinity Stress in Two Varieties of *Azolla* (*A. pinnata* and *A. filiculoides*). *Environ. Exp. Bot.* **2006,** *58*, 216–222.

Murthy, T. N. K.; Ashok, M.; Thirumalesh, T.; Umesh, B. U.; Nataraju, O. R. Effect of Partial Replacement of *Azolla* for Concentrate Supplement on Lactating Crossbred Cows. *Environ. Ecol.* **2013,** *31* (2), 415–417.

Parthasarathy, R.; Kadirvel, R.; Kathaperumal, V. *Azolla* as a Partial Replacement for Fish Meal in Broiler Rations. *Indian Vet. J.* **2002,** *79* (2), 144–146.

Rai, R. B.; Dhama, K.; Damodaran, T.; Ali H.; Rai S.; Singh B.; Bhatt, P. Evaluation of *Azolla* (*Azolla pinnata*) as a Poultry Feed and Its Role in Poverty Alleviation Among Landless People in Northern Plains of India. *Vet. Pract.* **2012,** *13* (2), 250–254.

Seulthrope, C. D. *The Biology of Aquatic Vascular Plants*; Edward Arnold Publisher Ltd.: London, 1967; pp 610–615.

Tamang, Y.; Samanta, G. Feeding Value of *Azolla* (*Azolla pinnata*) an Aquatic Fern in Black Bengal Goats. *Indian J. Anim. Sci.* **1993,** *63* (2), 188–191

Wagner, G. M. *Azolla*: A Review of Its Biology and Utilization. *Bot. Rev.* **1997,** *63* (1), 1–26.

CHAPTER 26

DISEASE MANAGEMENT OF NONLEGUMINOUS AND NONGRAMINACEOUS FORAGES

DIGANGGANA TALUKDAR[1] and UTPAL DEY[2*]

[1]*Department of Plant Pathology and Microbiology, College of Horticulture, Central Agricultural University, Ranipool 737135, Sikkim, India*

[2]*Division of Crop Production, ICAR Research Complex for NEH Region, Umiam 793103, Meghalaya, India*

[*]*Corresponding author. E-mail: utpaldey86@gmail.com*

ABSTRACT

Forage crops are primarily used as livestock feed, but it has the capacity to conserve and reclaim the soil thus benefitting in both conventional and applied agriculture. Nearly 200 plant species have been known as forage out of which some nonleguminous and nongraminaceous forage crops which include sunflower, *Brassica* spp., carrot, turnip, and *Amaranthus* are potential crops having tremendous scope as forage. However, these crops are prone to various biotic stressors, and among biotic stressors, several diseases caused by numerous fungus, bacteria, and virus are posing major threat in the growth and production of the crops. So, the utmost requirement for the proper management practices need to be enacted which includes integrated disease management principles such as use of plant extracts, biological agents, induced systemic resistance, seed treatment, cultural management, and chemical treatment such as fungicides and bactericides.

26.1 INTRODUCTION

Nonleguminous and nongraminaceous forage crops include sunflower, *Brassica* spp., carrot, turnip, and *Amaranthus*. Nonleguminous crops are those field crops that do not produce fruits as pods and do not have any nodules to fix atmospheric nitrogen. Nongraminaceous crops mean those crops which do not have woody hollow jointed stems and long narrow leaves. The term forage crops means those plant species which are grown for livestock feed and fodder as well as land conservation and reclamation of the soil. The vegetative portions of the plant, mainly leaves and stems, are consumed by the livestock. Thus, out of these four groups of crops, in this chapter, we have discussed about five very important crops that is sunflower, *Brassica* spp. (rapeseed and mustard), turnip, carrot, and *Amaranthus* which have lots of potential for being forage crop according to our purpose of utilization.

26.2 SUNFLOWER DISEASES

This crop is used for its edible oil and edible fruits (sunflower seeds). Apart from these uses, this sunflower species is also used as bird's feed, as livestock forage (as a meal or a silage plant), and in some industrial applications. It is the state flower of Kansas, US state, and one of the city flowers of Japan (Putnam et al., 2017).

The major diseases include Alternaria blight, rust, downy mildew, verticillium wilt, sclerotinia stalk and head rot, *Phoma* black stem, and leaf spot.

26.2.1 ALTERNARIA BLIGHT AND ITS MANAGEMENT

Patil et al. (1992) tested eight fungicides by the paper disc method against *Alternaria helianthi* and found ziram as the most effective at all the concentrations tested (0.1%, 0.2%, and 0.3%) followed by mancozeb and copper oxychloride. Wadiphasme et al. (1994) tested six nonsystemic and three systemic fungicides in vitro against *A. helianthi* by poison food technique. They found that Dithane M-45 was the most effective followed by Dithane Z-78. Kolte et al. (2000) reported that mancozeb at 0.3%, when sprayed four times at an interval of 7–10 days on sunflower, gave a good control of *A. helianthi*. The fungicides ziram and mancozeb were found highly effective in controlling *A. helianthi* in glass house and field trial (Bhaskaran and Kandaswamy, 1979). Rao (2006) reported a combi product iprodione + carbendazim @ 0.2% as effective fungicide for the management of Alternaria blight of sunflower.

Rao (2006) found neem leaf extract, neem kernel extract, and *Allium sativum* bulb extract as effective botanicals against *A. helianthi*.

The applications of *Pseudomonas* strains 679-2 to tomato and lucerne reduced the severity of leaf spot disease caused by *Alternaria solani* (Casida and Lokezic, 1992). Rao (2006) found *Pseudomonas fluorescens* and *Trichoderma harzianum* as effective bioagents in reducing the mycelial growth of *A. helianthi*.

Use of induced systemic resistance is a modern concept of plant disease management. Certain chemicals such as salicylic acid and jasmonic acid and organics such as cow urine and bioagents such as *Pseudomonas* sp. are known to induce systemic resistance when applied exogenously in small quantities (Mesta, 2006).

Mesta (2006) found that among the seed-dressing fungicides, iprodione + carbendazim along with Captan (both at concentration of 0.3 and 0.2) were found superior as they recorded lower percent infection and higher germination percentage and vigor index.

26.2.2 VERTICILLIUM WILT AND ITS MANAGEMENT

Rotation with nonsusceptible crops reduces the population of sclerotia in the soils. Good farm hygiene is preferred

26.2.3 DOWNY MILDEW AND ITS MANAGEMENT

It includes deep summer plowing, clean cultivation, and field sanitation. Avoid excessive irrigation, and remove infected plants

Use metalaxyl at 6 g/kg seed as seed treatment. Spray Ridomil MZ 72 WP @ 3 g/L at 20, 40, and 60 days after sowing.

26.2.4 POWDERY MILDEW AND ITS MANAGEMENT

Field sanitation, early varieties should be preferred. Spray with difenoconazole 25 EC @ 1 mL/L at initial stage and 15 days of first spray.

26.2.5 RUST AND ITS MANAGEMENT

Crop sanitation is done by spraying zineb 80 WP—2 g/L and mancozeb 75 WP—2 g/L. Spray these two fungicides at 15 days interval for two to three times (Shantamma and Patil, 2014).

26.3 RAPESEED AND MUSTARD DISEASES

Rapeseed and mustard are known as one of the most important oilseed crops worldwide. They belong to group *Brassica* spp., that is, rapeseed as *Brassica rapa* and mustard as *Brassica juncea*. In general, there are four types of rapeseed and mustard, rai or Indian mustard (*B. juncea*); yellow and brown sarson (*Brassica campestris*); toria or rapeseed (*B. rapa*); and tara (*B. napus*).

Apart from producing oil, rapeseed and mustard are used as cover crops, for rapid growth, great biomass production, and nutrient scavenging ability. Most of the *Brassica* species release chemical compounds that may be toxic to soil-borne pathogens and pest and weeds. Thus, this crop can be used for erosion control, as pest, disease, and nematode and weed management.

One of the major concerns in enhancing the yield of rapeseed and mustard is the incidence of diseases, which are causing damage to crop at different stages and responsible for huge yield losses to an extent ranging from 10% to 90%. In present article, important pests of mustard and their integrated management has been described in detail.

26.3.1 ALTERNARIA BLIGHT (Alternaria brassicae) AND ITS MANAGEMENT

Alternaria blight disease caused by *Alternaria* spp. has been reported from all the continents of the world and is one among the important diseases of the crop in India causing up to 47% yield losses (Kolte, 1985). Saharan

(1992) and Kolte (2002) reported that Alternaria blight sometimes causes more severe losses (up to 70%) in rapeseed (*B. campestris*). The blight also reduces seed size and impairs seed color and oil content (Kaushik et al., 1984).

Three systemic fungicides: Topsin-M (thiophanate methyl, 70% WP), Ridomil MZ (mancozeb, 64% + metalaxyl, 8% WP), and Bavistin (carbendazim, 50% WP) alone and in combination with four nonsystemic fungicides Captaf (Captan, 50% WP), Indofil M-45 (mancozeb, 75% WP), Indofil Z-78 (zineb, 75% WP), and Thiram (Thiram, 75% WP) were evaluated both in vitro and in vivo for their effectiveness to manage Alternaria blight of rapeseed mustard caused by *Alternaria brassicae* (Khan et al., 2007).

Meena et al. (2011) reported that mancozeb recorded the lowest mean severity (leaf: 33.1%; pod: 26.3%) of Alternaria blight with efficacy of garlic bulb extract alone (leaf = 34.4%; pod = 27.3%) or in combination with cow urine (leaf = 34.2%; pod = 28.6%) being statistically at par with the recommended chemical fungicide.

Meena et al. (2010) found that there are certain biological agents responsible for controlling Alternaria blight of rapeseed and mustard. Spray of soil isolates of *Trichoderma viride* at 45 and 75 days after sowing could manage Alternaria blight of Indian mustard (*B. juncea*) as effectively as mancozeb (Meena et al., 2004), which have been confirmed later in multilocation trials.

Bulb extract of *A. sativum* has been reported to effectively manage Alternaria blight of Indian mustard (Kolte, 2002).

Early sowing (Meena et al., 2002) of well-stored clean-certified seeds after deep plowing, clean cultivation, timely weeding and maintenance of optimum plant population, avoidance of irrigation at flowering, and pod formation stages may help to manage the disease.

Soil application of K as basal has been found to check Alternaria blight disease in mustard. Studies on variability at pathogenic and genetic level in *A. brassicae* could enable easier development of disease-resistant material.

Identification of signal molecules for induced resistance, development of bioformulations and disease forecasting techniques based on epidemiological findings will enable trigger newer strategies for environment-friendly disease management for providing safer Alternaria blight-free production of Indian mustard (Meena et al., 2002).

26.3.2 DOWNY MILDEW (*Peronospora brassicae*) AND ITS MANAGEMENT

In the disease affected plants, yellow, irregular spots appear on the upper surface of the leaves and white growth is visible on the under surface opposite to spots. If the attack is severe, inflorescence is also affected. The affected inflorescence is malformed, twisted, and covered with a white powder. No pods are produced on such inflorescence.

Use healthy seeds for sowing. Spray the crop with 0.2% zineb or 0.1% Karathane as soon as the symptoms are noticed and repeat the spray two to three times at 10 days interval (Ahalwat, 2008).

26.3.3 WHITE BLISTER (*Albugo candida*) AND ITS MANAGEMENT

This disease can be a serious menace if it occurs along with downy mildew. The disease is characterized by white raised blisters on leaves, stem, petiole, and floral parts. These blisters burst and liberate a white powder. There is much deformity of the floral parts. Flowers get malformed and become sterile.

Use healthy seeds for sowing. Spray the crop with 0.2% zineb or Difolatan as soon as the symptoms are noticed and repeat the spray if needed at 10 days interval. Keep the field free from weeds.

26.4 TURNIP DISEASES

Turnip, (scientific name: *Brassica rapa*), is an herbaceous annual or biennial plant in the family Brassicaceae which is grown for consumption of edible roots and leaves. The plant possesses erect stems with 8–12 leaves forming a crown. The leaves are light green in color, hairy, and thin. The plant produces light yellow flowers which are clustered at the top of a raceme and are often extended above the terminal buds. The leaves can reach 30.5–35.5 cm in length, while the branching flower stems can reach 30.5–91.5 cm. The taproot of the plant is a bulbous tuber, almost faultlessly round, which is usually a mixture of purple, white, and yellow. Turnip is usually grown as an annual and harvested after one growing season. Turnip may also be referred to as annual turnip and originated from Europe. Turnips are eaten as a vegetable after cooking. The shoots and leaves can be eaten fresh in salads or the entire plant can be used as forage for livestock. But despite of its numerous uses, it is susceptible to many diseases. There is proper need for the management of the disease. Turnip is beneficial as it can suppress weeds, grow at quick pace, scavenge soil nitrogen below root zone, and most importantly can be used as animal feed.

26.4.1 ALTERNARIA LEAF SPOT AND ITS MANAGEMENT

Causal organism is a fungi *Alternaria* spp. Small dark spots on leaves which turn brown to gray; lesions may be round or angular and may possess a purple–black margin; lesions may form concentric rings, become brittle and crack in center; dark brown elongated lesions may develop on stems and petioles. Plant only pathogen-free seeds; rotate crops; applications of appropriate fungicides control disease when present.

26.4.2 ANTHRACNOSE AND ITS MANAGEMENT

Causal organism is a fungi *Colletotrichum higginsianum*. Small circular or irregularly shaped dry spots which are gray to straw in color develop on leaves; a high number of spots may cause the leaf to die; lesions may coalesce to form large necrotic patches causing leaves to turn yellow and wilt; lesions may split or crack in dry centers; dry sunken spots on roots which enlarge and turn gray or brown. Fungus overwinters on leaf debris and on related weeds; disease emergence is favored by moist, warm conditions.

Control of disease depends on sanitary practices; treat seeds with hot water prior to planting, rotate crops, plant in an area with good soil drainage, and remove all cruciferous weeds which may act as a reservoir for the fungus. Fungicide application must begin with first appearance of the symptoms and continue at 7–10 days intervals as long as weather conditions favor disease development.

26.4.3 BLACK ROOT AND ITS MANAGEMENT

Causal organism is a fungi *Aphanomyces raphani*. Small black–blue areas appear on roots which expand and girdle taproot; roots become constricted at site of lesions; black discoloration extends into root. Fungus can survive in soil for prolonged periods of time. Control depends on crop rotation with non-brassica species.

26.4.4 BLACK ROT AND ITS MANAGEMENT

Causal organism is a bacteria *Xanthomonas campestris*. Seedlings develop wilted yellow to brown leaves and collapse; yellow, V-shaped lesions develop on mature leaf margins; dark rings can be found in the cross section of the stem. Disease emergence favors warm, wet conditions.

Primary control methods based on good sanitation; plant disease-free seeds, rotate crops every 2 years or less to non-brassica, and avoid sprinkler irrigation. A bactericide called Kocide DF @ 1–2 lb at 7–10 days interval can be applied at initiation of the symptoms.

26.4.5 CERCOSPORA LEAF SPOT (FROGEYE LEAF SPOT) AND ITS MANAGEMENT

Causal organism is a fungi *Cercospora brassicicola*. Angular or circular green to gray spots with brown borders appear on leaves; plant defoliation may occur in the case of a severe infestation. Disease emergence favors cool temperatures and wet weather.

Plant only certified disease-free seeds; avoid overhead irrigation; rotate crops to non-brassica species for 2–3 years; apply appropriate fungicide if disease emerges. Thiolux (sulfur) @ 3–10 lb applied at 10–14 days interval proved to be effective in reducing the disease incidence.

26.4.6 CLUBROOT AND ITS MANAGEMENT

Causal organism is fungi *Plasmodiophora brassicae*. Slow growing, stunted plants; yellowish leaves which wilt during day and rejuvenate in part at night; swollen, distorted roots; extensive gall formation. Can be difficult to distinguish from nematode damage; fungus can survive in soil for periods in excess of 10 years; can be spread by movement of contaminated soil and irrigation water to uninfected areas.

Once the pathogen is present in the soil it can survive for many years, elimination of the pathogen is economically unfeasible; rotating crops generally does not provide effective control; plant only certified seeds and avoid field grown transplants unless produced in a fumigated bed; applying lime to the soil can reduce fungus sporulation.

26.4.7 DOWNY MILDEW AND ITS MANAGEMENT

Causal organism is a fungi *Peronospora parasitica*. Irregular yellow patches on leaves which turn light brown; fluffy gray growth on the undersides of the leaves. Disease emergence favors cool temperatures; disease spreads quickly in wet conditions.

Remove all crop debris after harvest; rotate with non-brassicas; application of appropriate fungicides may be required if symptoms of disease are present. Fungicides such as Aliette @ 2–5 lb at 7–21 days or Koicide DF @ 1–2 lb at 7–10 days' interval must be applied.

Crop rotation is done every 1–3 years with noncruciferous crops. Destroy and burn the weeds. Destroy the diseased plants by shredding and disking or tilling as soon as harvest is completed. Control the aphids that act as vector for many diseases. Avoid the application of high levels of nitrogen fertilizer; it should be done on the soil requirement. Maintain well drained soil. Plant must be sown in raised bed. Planting must be done in such a way that it receives the early morning sun.

26.5 CARROT DISEASES

Carrot, *Daucus carota*, is an edible, biennial herb in the family Apiaceae grown for its edible root. The carrot plant produces a rosette of 8–12 leaves aboveground and a fleshy conical taproot below ground. The plant produces small (2 mm) flowers which are white, red, or purple in color. The root can grow to between 5 and 50-cm-long (2.0–20 in.) and reach 5 cm (2.0 in.) in diameter. The foliage of the plant can reach a height of 150 cm (59.1 in.) when in flower. The carrot plant can be annual or biennial and may also be referred to as wild carrot. The plant is believed to have originated in Europe or the Western Mediterranean. Carrot roots are eaten as a vegetable and can be consumed fresh or cooked. Carrot juice is consumed as a beverage. The leaves of the plant can be used as feed for animals.

According to FAO (1994), carrot can be one of the components of forage crop. Carrot can be nutritious feed for all types of livestock. The leaves, stems, and roots are highly palatable to livestock (Fraser et al., 1970). It contains beta-carotene that helps in winter feed for dairy cattle to produce yellow coloring in butter and cream. Cattle can be fed 40–60 lb of carrots per day; pigs and poultry can feed 8 lb of carrot per meal.

26.5.1 ALTERNARIA LEAF BLIGHT AND ITS MANAGEMENT

Causal organism is fungi *Alternaria dauci*. It is characterized by green to brown water-soaked lesions on leaves which enlarge and turn dark brown or black. Later on these lesions may coalesce causing leaves to yellow and ultimately die. Lesions may also spread on petioles. The emergence of this disease is favored by wet foliage and warm weather, while rain and fog enhance the development of the disease; fungus survives in soil on crop debris but is killed when the debris decomposes (Davis and Raid, 2002). It is may be seed-borne and may spread on carrot seeds. It also survives in carrot debris and in the soil for several years.

Application of appropriate fungicides when first symptoms appear will result in decrease in this disease infection to many folds. Crop rotation is highly recommended to prevent buildup of the fungus in the soil. Deep tillage may provide some control by burying the inoculums of the fungus away from the carrot crown where most of the fungal spores harbor. Treating seeds with fungicide or hot water prior to planting is very helpful. Applications of gibberellic acid to carrot foliage promote upright growth and air circulation through canopy. Dusting seed with Thiram or iprodione can significantly reduce this seed-born disease (Maude and Bambridge, 1991; Stranberg, 1988).

26.5.2 BACTERIAL LEAF BLIGHT AND ITS MANAGEMENT

Causal organism is a bacterium *Xanthomonas campestris* pv. *carotae*. It is characterized by small, angular, yellow spots on leaves which enlarge into irregularly shaped, brown, water-soaked lesions with a yellow halo surrounding. Centers of lesions dry out and become brittle. Leaves may become curled or distorted. Flower stalks may develop elongated lesions that exude a bacterial ooze. Infected umbels may be blighted with the pace of time amalgamated with favorable and congenial environment. Bacteria can be spread by splashing irrigation water or rain or on contaminated equipment.

It can be achieved by using pathogen-free seed, avoiding sprinkler irrigation, and applying appropriate bactericides if accessible. Sow or plant *Xanthomonas*-indexed seed or treat the seeds in hot water (52°C for 25 min). Use furrow or drip irrigation rather than sprinklers as mentioned above. Turn underneath carrot residue to accelerate decomposition. Avoid continuous carrot monoculture by using a 2–3-year crop rotation scheme.

Cultural controls: hot water dips, and sprays of certain copper sulfate formulations are acceptable for use on organically grown produce

26.5.3 BLACK ROT AND ITS MANAGEMENT

Causal organism is a fungi *Alternaria radicina*. The most typical symptom is damping-off of seedlings. Root and crown necrosis can be observed. There is appearance of blighted foliage and lower portion of petioles turned black and necrotic. Black ring around petiole and sunken lesions on taproot were seen. Disease is spread through infected seed and can survive in soil for up to 8 years.

Black rot is difficult to control and can survive in the soil for longer periods of time. Therefore, it is recommended to practice long crop rotations. Plowing crop residue into soil immediately after harvest is helpful in reducing the infection. Use of plant-resistant varieties and pathogen-free seeds and treating seeds with hot water prior to planting are effective in managing the infection.

26.5.4 CAVITY SPOT AND ITS MANAGEMENT

Causal organism is a fungi *Pythium* spp. It is identified by sunken, elliptical, gray lesions across the root. Outer layer of root ruptures and develops dark, elongated lesions. Small vertical cracks may form on the cavities. Fungi can persist in soil for several years and disease outbreaks are associated with wet soils. Flooded soil increases the number of cavities formed in the skin of the vegetable.

Some cultural practices can control this disease. Avoid planting in fields or areas which had record of carrot having carrot spot previously. Do not overfertilize plants with nitrogen. Applications of appropriate fungicide can provide adequate control.

26.5.5 CERCOSPORA LEAF BLIGHT AND ITS MANAGEMENT

Causal organism is a fungi *Cercospora carotae*. It is characterized by small, necrotic flecks on leaves which develop a chlorotic halo around the diseased part and expand into tan brown necrotic spots. The lesions coalesce and cause leaves to shrivel, curl, and die. Disease can be introduced through infested seed and spread by wind or water splash. Symptoms usually occur first on the younger foliage.

It can be achieved by planting only pathogen-free seeds. Crop rotation is must. Plow crop debris into soil after harvest. Apply appropriate fungicidal sprays as and when required.

26.5.6 COTTONY ROT (SCLEROTINIA ROT) AND ITS MANAGEMENT

Causal organism is a fungi *Sclerotinia sclerotiorum*. Small, water-soaked, soft lesions on crown and roots are observed. There is appearance of white fluffy fungal growth all over affected tissues. Soft and decaying tissue develops in the affected area. Fungus can survive in soil for up to 10 years.

Cultural practices play an important role in the control of cottony rot as there are no resistant carrot varieties developed till date. In carrot fields, the use of drip irrigation 5–8 cm below the soil surface can provide a good control. Deep plowing of soil and trimming back carrot foliage to promote air circulation can also be useful. Application of fungicides may be dispensable in periods of extended cool and damp weather conditions.

26.5.7 DAMPING-OFF AND ITS MANAGEMENT

Causal organism is a fungi *Pythium* spp. It leads to production of soft and rotting seeds which fail to germinate. Rapid death of seedlings prior to emergence from soil is the most common phenomenon. Collapsing of seedlings after they have emerged from the soil is caused by water-soaked reddish lesions girdling the stem at the soil line. Damping-off diseases favor conditions which slow seed germination. Fungi can be spread in water, contaminated soil, or on equipment.

Avoid planting carrots in poorly draining, cool, wet soil; planting in raised beds will help with soil drainage; plant high quality seeds that germinate quickly; treat seeds with fungicide prior to planting to eliminate fungal pathogens.

26.5.8 DOWNY MILDEW AND ITS MANAGEMENT

Causal organism is a fungi *Peronospora umbelliferum*. Yellow spots on upper surface of leaves are observed. White fluffy growth on underside of leaves is found. Color of the lesions become darker as it matures. Disease affects young and tender leaves. Emergence of the disease is favored by prolonged leaf wetness.

Planting pathogen-free seeds is recommended. Overcrowded planting is strictly avoided. Crop rotation with nonumbelliferous varieties reduces the disease.

26.5.9 POWDERY MILDEW AND ITS MANAGEMENT

Causal organism is a fungi *Erysiphe heraclei*. Powdery growth is spread on leaves, petioles, flowers stalks, and bracts leaves become chlorotic; severe infections can cause flowers to become distorted. Fungus can spread long distances in air; disease emergence is favored by high humidity and moderate temperatures. Plants are more susceptible when grown in shady locations or stressed by drought. Susceptibility also increases with plant age, most notably starting 7 weeks after seeding.

Monitoring of the plants from the early stage is necessary as powdery mildew is very difficult to see on leaves in early stage. Plant tolerant varieties if available. Avoid excess fertilization. Protective fungicide applications (e.g., chlorothalonil) provide adequate protection. Sulfur application can be used in infection that occurs early in season.

26.5.10 ROOT KNOT NEMATODES (STUBBY ROOT NEMATODES AND NEEDLE NEMATODES) AND THEIR MANAGEMENT

Causal organisms are nematodes *Meloidogyne* spp.; *Paratrichodorus* spp.; and *Longidorus africanus*. Forked, distorted, or stunted taproots are observed. There is reduction in crop stand and yield. Root knot nematodes are most damaging to carrot.

Leaving land to fallow when not planting can be effective at reducing nematode numbers. Soil solarizing for 4–6 weeks period to a depth of 6 in. can temporarily reduce nematode populations. New carrot varieties are currently being developed that are resistant to nematodes.

The root knot nematodes *Meloidogyne incognita*, *Meloidogyne javanica*, and *Meloidogyne arenaria* do not penetrate roots at soil temperatures below 59–64°F. Therefore, planting should be done when soil temperatures are below this level. Galls and egg masses of the nematodes in secondary roots of the carrots could significantly reduce in broccoli-amended soil with solarization and *Trichoderma* inoculation as reported by Pedroche et al. (2009).

26.5.11 SOFT ROT AND ITS MANAGEMENT

Causal organism is a mixture of several species of bacteria, namely, *Erwinia carotovora*, *Erwinia chrysanthemi*, and *Pseudomonas marginalis*. Sunken dull orange lesions are observed on taproot which causes tissue to collapse and become soft. Bacteria thrive in oxygen-depleted plant tissue. Disease emergence requires long periods of water-saturated soil. Bacteria enter plants through natural openings and wounds.

Control relies on the avoidance of conditions conducive to bacterial infection: plant carrots in well-draining soils; allow plants to dry before irrigating again; avoid wounding plants during harvest to prevent postharvest development of disease; disinfect all equipment regularly.

26.5.12 TURNIP MOSAIC AND ITS MANAGEMENT

Causal organism is a virus Turnip mosaic virus. Yellow and green mosaic patterns on leaves, necrotic areas on leaves, vein clearing and chlorosis may occur in older leaves, black spots and brown necrotic streaks on stem, stunted plant growth, and reduced yield. It is transmitted by many species of aphids such as cabbage aphid and peach aphids. Use of reflective mulches may repel the aphid on feeding, application of appropriate fungicides.

26.6 AMARANTHUS DISEASES

Amaranth (*Amaranthus* spp.) is a dicotyledonous plant and the name is given to a group of approximately 70 species of annual or short-lived perennial plants in the genus *Amaranthus* including several species of aggressive edible weeds native to the United States such as *Amaranthus retroflexus* (redroot pigweed). Amaranths are branching broad-leaved plants with egg-shaped or rhombic leaves which may be smooth or covered in tiny hairs. The leaves have prominent veins, can be green or red in color, and have long petioles. The plants produce single flowers on terminal spikes which are typically red to purple in color. Amaranths can reach up to 2.5 m (6.6 ft) in height and are usually grown as annuals, harvested after one growing season. Amaranths may also be referred to as Chinese spinach and their origin is unclear due to their worldwide distribution. Amaranth leaves and stems are commonly eaten after cooking in a manner similar to spinach. There are four main species which are cultivated as vegetables; *Amaranthus cruentus*, *A maranthus blitum*, *Amaranthus dubius*, and *A maranthus tricolor.* Several species, such as, *Amaranthus caudentis*, *A. cruentus*, and *A maranthus hypochondriacus* are grown as a grain crop in places such as Mexico, Nepal, and India and are used to produce cereals and snack.

This crop has a huge potential as forage crop (Leukebandara et al., 2015). Therefore, introduction of this plant as a new forage crop for livestock production is highly acceptable and beneficial, on the basis of nutritional significance also making up an opportunity for diversification of animal feeding systems. In Sri Lanka, experiment was conducted by Leukebandara et al. (2015) to see animal performance trials, digestibility, and palatability studies with feeding Amaranth, in different livestock species, in different farming systems of Sri Lanka and found lots of potential qualities for this crop as forage. *Amaranthus* supplies a substantial part of the protein, minerals, and vitamins in the diet, thus increased the scope to cultivate more vehemently and domesticate.

26.6.1 ANTHRACNOSE AND ITS MANAGEMENT

Causal organism is a fungi *Colletotrichum gloeosporioides.* Necrotic lesions develop on leaves; dieback of leaves and branches. Avoid damaging plants that create wounds for pathogen to enter. Use plant-resistant varieties.

26.6.2 DAMPING-OFF AND ITS MANAGEMENT

Causal is involvement of two fungi together, that is, *Rhizoctonia* spp. and *Pythium* spp. The most common disease is poor germination and seedling collapse. Brown to black lesions girdling stem close to soil line is observed. Seedlings fail to emerge from soil. Disease emergence favors in wet soils.

Use of disease-free seeds, avoiding planting of seeds too deeply, and avoiding dense planting so as to promote air circulation around seedlings reduces the infection. Care should be taken not to overwater the plants.

26.6.3 WET ROT (CHOANEPHORA ROT) AND ITS MANAGEMENT

Causal organism is a fungus *Choanephora cucurbitarum*. Water-soaked lesions are formed on the stems. The lesions have hairy appearance (silk-like threads) due to presence of fungal spores. It may cause heavy defoliation during rainy seasons. Fungus mainly attacks plants that have been damaged by insects or by mechanical means; spread by air currents and via infected seed; disease emergence favors warm, moist conditions (Seymour, 2013).

Plant varieties resistant to disease must be used. Only use certified seeds recommended. Avoid dense planting to allow sufficient aeration. Treat disease with copper fungicides if it emerges and practice good field sanitation.

KEYWORDS

- **nongraminaceous forage**
- **nonleguminous forage**
- **integrated diseases management**

REFERENCES

Bhaskaran, R.; Kandaswamy, T. K. Evaluation of Fungicides for the Control of Alternaria Leaf Spot of Sunflower. *Indian J. Agric. Sci.* **1979,** *49*, 480–482.

Casida, L. E.; Lokezic, F. L. Control of Leaf Spot Disease of Alfalfa Tomato with Application of the Bacterial Predators, *Pseudomonas* Strains 679-2. *Plant Dis.* **1992,** *76*, 1217–1220.

Davis, R. M.; Raid, R. M. *Compendium of Umbelliferous Crop Diseases*; American Psychopathological Society Press: Crown Plaza, New Work, 2002.

Fraser, S.; Gilmore J. W.; Clark C. F. Culture Varieties of Roots for Stock Feeding. *Cornell University Agriculture Experiment Station Bulletin*, 1970, p 264.

Kaushik, C. D.; Saharan, G. S.; Kaushik, J. C. Magnitude of Loss in Yield and Management of Alternaria Blight in Rapeseed-mustard. *Indian Phytopathol.* **1984,** *37*, 398.

Khan, M. M.; Khan, R. U.; Mohiddin, F. A. Studies on the Cost-effective Management of Alternaria Blight of Rapeseed-mustard (*Brassica* spp.). *Phytopathol. Mediterr.* **2007,** *46*, 201–206.

Kolte S. J. *Diseases and Their Management in Oilseed Crops, New Paradigm in Oilseeds and Oils: Research and Development Needs*; Rai, M., Singh, H., Hegdeed, D. M., Eds.; Indian Society of Oilseeds Research: Hyderabad, India, 2002; pp 264–252.

Kolte, S. J.; Awasthi, R. P.; Vishwanath. Divya Mustard: A Useful Source to Create Alternaria Black Spot Tolerant Dwarf Varieties of Oilseed Brassicas. *Plant Varieties Seeds* **2000,** *13*, 107–111.

Leukebandara, I. K.; Premaratne S.; Peiris, B. L. Nutritive Quality of Thampala (*Amaranthus* spp.) as a Forage Crop in Sri Lanka. *Trop. Agric. Res.* **2015,** *26* (4), 626–631.

Maude, R. B.; Bambridge, J. M. Evaluation of Seeds Treatments Against *Alternaria dauci* (Leaf Blight) of Naturally Infected Carrot Seeds. *Ann. Appl. Biol.* **1991,** *118*, 30–31.

Meena, P. D.; Chattopadhyay, C.; Singh, F.; Singh, B.; Gupta, A. Yield Loss in Indian Mustard due to White Rust and Effect of Some Cultural Practices on Alternaria Blight and White Rust Severity. *Brassica* **2002,** *4*, 18–26.

Meena, P. D.; Meena, R. L.; Chattopadhyay, C.; Kumar, A. Identification of Critical Stage for Disease Development and Biocontrol of Alternaria Blight of Indian Mustard (*Brassica juncea*). *J. Phytopathol.* **2004,** *152*, 204–209.

Meena, P. D.; Awasthi, R. P.; Chattopadhyay, C.; Kolte, S. J.; Kumar A. Alternaria Blight: A Chronic Disease in Rapeseed-mustard. *J. Oilseed Brassica* **2010,** *1* (1), 1–11.

Meena P. D.; Chattopadhyay, C.; Kumar A.; Awasthi R. P.; Singh, R.; Kaur, S.; Thomas, L.; Goyal, P.; Chand, P. Comparative Study on the Effect of Chemicals on Alternaria Blight in Indian Mustard: A Multi-location Study in India. *J. Environ. Biol.* **2011,** *32*, 375–379.

Mesta, R. K. Epidemiology and Management of Alternaria Blight of Sunflower Caused by *Alternaria helianthi* (hansf.) Tubaki and Nishihara. Ph.D. (Agri) Thesis, University of Agricultural Sciences, Dharwad, 2006; p 89.

Patil, M. K.; Kulkarni, S.; Hegde, Y. In Vitro Bioassay of Fungicides Against Leaf Spot of Safflower. *Curr. Res.* **1992,** *21*, 60.

Pedroche, N. B.; Villanueva, L. M.; De Waele, D. Management of Root-knot Nematode, *Meloidogyne incognita* in Carrot. *Commun. Agric. Appl. Biol. Sci.* **2009,** *74* (2), 605–615.

Putnam, D. H.; Oplinger, E. S.; Hicks, D. R.; Durgan, B. R.; Noetzel, D. M; Meronuck, R. A.; Doll, J. D.; Schulte, E. E. Sunflower. In *Corn Agronomy*; University of Wisconsin: Madison, WI, 2017; pp 262–1390. [1575 Linden Drive—Agronomy, Madison WI, 53706 (608)].

Rao, M. S. L. Studies on Seed Borne Fungal Diseases of Sunflower and Their Management. Ph.D. Thesis, University of Agricultural Sciences, Dharwad, 2006; pp 55.

Seymour, T. Foraging New England. In *Falcon Guides*, 2nd ed.; Morris Book Publishing, LLC: Kearney, NE 68848-2110, 2013.

Stranberg, J. O. A Selective Medium for the Detection of *Alternaria dauci* and *Alternaria radicina*. *Phytoparasitica.* **2002,** *30*, 269–284.

Wadiphasme, S. S.; Ingole, D. V.; Fulzele, G. R. Evaluation of Fungicides *In Vitro* Against *Alternaria helianthi* Inciting Leaf Blight of Sunflower. *J. Maharashtra Agric. Univ.* **1994,** *19*, 361–363.

INDEX

A

Amaranthus (pigweed)
  botanical classification, 225–226
  botanical name, 226
  climatic requirement, 228
  common name, 226
  crop mixture, 231
  description, 227
  disease management, 233
  harvesting, 233
  insect-pest management, 233
  nutrient management, 232
  nutritive value, 234
  plant characteristics, 227–228
  seed
    inoculation, 231
    production, 234
    rate, 231
    sowing method, 231
    treatment, 231
  soil and preparation, 228
  sowing time, 231
  special features (toxicities), 235–236
  utilization, 235
  varieties, 229
  water management, 232
  weed management, 233
  yield, 234
Animal health issues, 203
  goiter (enlarged thyroid), 204
  kale anemia, 204–205
  nitrate poisoning, 204
  photosensitization, 204
Anjan grass (African foxtail grass), 55
  botanical classification, 56
  botanical name, 56
  climatic requirement, 58
  common name, 56
  cropping systems, 60
  disease management, 61
  geographical distribution, 57–58
  green fodder yield, 61
  harvesting, 61
  insect-pest, 61
  interculture operation, 60
  nutrient management (manures and fertilizers), 60
  nutritive value, 62
  plant characteristics, 57
  seed
    production, 62
    rate, 59
    sowing method, 59
  soil and preparation, 58
  sowing/planting time, 59
  spacing, 59–60
  special features (toxicities), 63
  utilization, 62
  varieties, 58–59
  water management, 60
  weed management, 60
Azolla, unconventional forage crop, 309
  botanical classification, 310
  botanical name, 310
  common name, 310
  cultivable species of, 313
  cultivation of, 313
  environmental requirements, 312
  geographical distribution, 311
  harvesting and yield, 314
  intercultural operation, 313
  morphology, 311
  natural habitat, 312
  origin, 311
  utilization
    biofertilizer in rice cultivation, 315–316
    nutritional supplement for livestock, 316–317
    uses, 317

B

Brassicas, 187
animal health issues, 203
goiter (enlarged thyroid), 204
kale anemia, 204–205
nitrate poisoning, 204
photosensitization, 204
botanical classification, 188
botanical name, 188
climatic requirement, 193
soil requirement, 194
common name, 188
diseases management
alternaria leaf spot, 201
black leg, 201
club root, 200
damping off, 201
white rust, 201
fertilizer requirement, 197
field preparation, 194
forage brassicas, types, 189
Chinese cabbage, 192
hybrids, 191
kale, 191
leafy turnips, 191
rape, 191
swedes, 192
turnip, 191–192
grazing, 202–203
harvesting and yield, 201–202
insect-pests management
APHID, 199–200
cabbage butterfly, 200
cruciferous leaf webber, 199
diamond back moth, 199
tobacco caterpillar, 200
origin and distribution, 192–193
planting time, 196
seed
treatments, 195
sowing method, 194–195
varieties, 196
water management, 198
weed management, 198

C

Canary grass (Harding grass)
botanical classification, 47–48
climatic requirements, 49
common name, 48
crop mixtures, 50
depth of sowing, 50
green fodder yield, 52
harvesting, 52
intercultural operation, 51
irrigation and drainage, 51
land preparation, 49–50
manures and fertilizers, 51
nutritive value, 52
origin and geographical distribution, 49
pest and diseases, 52
plant characteristics, 49
scientific name, 48
seed rate, 50
sowing, 50
spacing, 50
toxicities, 52
utilization, 45, 53
varieties, 51
weed management, 51

D

Diseases management
alternaria leaf spot, 201
black leg, 201
club root, 200
damping off, 201
white rust, 201
Doob grass (Bermuda grass), 21
botanical classification, 22
botanical name, 22
climatic requirements, 23–24
common name, 22
cropping systems, 26
description, 23
disease management, 27
green fodder yield, 28
harvesting, 28
insect-pest, 27
interculture operation, 27

nutrient management (manures and fertilizers), 26–27
nutritive value, 28
origin and distribution, 23
plant characteristics, 23
seed
inoculation, 25–26
production, 28
rate, 26
sowing method, 26
treatment, 25–26
soil and its preparation, 24
sowing/planting time, 25
spacing, 26
special features (toxicities), 29
utilizations, 29
varieties, 24–25
water management, 27
weed management, 27

F

Fenugreek (Greek clover)
botanical classification, 121–122
botanical name, 122
climatic requirement, 123
common name, 122
cropping systems, 125
disease management, 126
harvesting, 127
insect-pest, 126
interculture operation, 126
nutrient management (manures and fertilizers), 125–126
nutritive value, 127
origin and distribution, 122–123
plant characteristics, 123
production, 123
seed
inoculation, 124
production, 127
rate, 124
sowing method, 124
treatment, 124
soil and preparation, 123–124
sowing time, 124
spacing, 125
toxicities, 127
utilization, 128
varieties, 124
water management, 126
weed management, 126
yield, 127
Forage brassicas, types, 189
Chinese cabbage, 192
hybrids, 191
kale, 191
leafy turnips, 191
rape, 191
swedes, 192
turnip, 191–192

G

Gajar (carrot)
botanical classification, 215–216
botanical name, 216
climatic requirement, 217
common name, 216
crop mixture, 219
disease management, 221
harvesting, 221
insect-pest management, 221
nutrient management, 219
nutritive value, 222
origin and distribution, 216–217
plant characteristics, 217
seed
production, 221
rate, 218
sowing method, 218
soil and preparation, 217–218
sowing time, 218
spacing, 218
utilization, 222
varieties, 218
water management, 219–220
weed management, 220
yield, 221
Gliricidia (quickstick)
botanical classification, 251
botanical name, 252
climatic requirement, 253
common name, 252

compatibility, 256
description, 252
disease management, 255
harvesting, 255
insect-pest management, 255
intercropping, 254
nutritive value, 255
origin and distribution, 253
plant characteristics, 252
seed
rate, 254
sowing method, 254
treatment, 254
soil, 253
sowing time, 253
special features, 256
utilization, 255–256
water management, 254
weed management, 254
yield, 255
Grass pea (Indian vetch)
botanical classification, 77–78
botanical name, 78
climatic requirement, 81
common name, 78
cropping systems, 83
disease, 84
harvesting, 84–85
insect-pest management, 84
lathyrus, 87–88
nutrient management (manures and fertilizers), 83
nutritive value, 85–86
origin and distribution, 80–81
plant characteristics, 79–80
seed
rate, 82
sowing method, 82
treatment and inoculation, 83
soil and preparation, 81
sowing time, 82
toxicities, 86–87
utilization, 86
varieties, 81–82
water management, 83–84
weed management, 84
yielding, 85

I

Insect-pest management
beetles, 275
locust, 275
shoot gall maker, 276
termite, 276
Insect-pests management
APHID, 199–200
cabbage butterfly, 200
cruciferous leaf webber, 199
diamond back moth, 199
tobacco caterpillar, 200

K

Khejri (prosopis), 259
botanical classification, 260
botanical description, 261
botanical name, 260
climatic requirement, 263–264
common name, 260
crop regulation, 273–274
cropping systems, 274–275
disease management, 276
germplasm conservation, 264
harvesting, 276–277
insect-pest management
beetles, 275
locust, 275
shoot gall maker, 276
termite, 276
intercropping, 271–272
maturity indices, 276–277
nutrient management, 269
nutritive value
fruits, 280
leaves, 280
orchard establishment, 267–269
origin and distribution, 262
plant characteristics, 262–263
seed propagation, 265
soil, 263
training and pruning, 272–273
utilization
bark, 279
flower, 280
green pods, 277–278

gum, 279
leaves fodder, 278–279
ripe pods, 278
wood, 279
worship of, 280
varietal wealth, 264–265
vegetative propagation
air layering, 266
budding, 266
cutting, 265–266
micropropagation, 266–267
root suckers, 265
rootstock, 267
water management, 269–270
weed management, 271
yield, 276–277

M

Marvel grass (diaz blue stem)
botanical classification, 31
botanical name, 32
climatic requirements, 33
common name, 32
compatibility, 36
crop mixture, 34
harvesting, 35
insect-pest and diseases, 35
intercropping, 34
nutrient management, 34
nutritive value, 35–36
origin and distribution, 32–33
plant characteristics, 33
seed
production, 35
rate, 34
sowing method, 34
treatment, 34
soil and its preparation, 33
sowing time, 34
synonyms, 32
toxicity, 36
utilizations, 36
varieties, 33
water management, 35
weed management, 35
yield, 35
Moth bean (dew bean)
acreage and production, 94
botanical classification, 91–92
botanical name, 92
climatic requirement, 94
common name, 92
cropping systems, 96
description, 93
disease management, 97
harvesting, 98
insect-pest, 97
intercultural operation, 97
manures and fertilizers, 97
nutritive value, 99
origin and distribution, 93
seed
inoculation, 96
production, 98
rate, 96
sowing method, 96
treatment, 96
soil and preparation, 94
sowing time, 96
utilization, 99
varieties, 94–95
water management, 97
weed management, 97
yield, 98
Mung bean (green gram)
botanical classification, 139–140
botanical name, 140
climatic requirement, 143
common name, 140
cropping systems, 146
diseases management, 148–149
harvesting, 149
insect-pests management, 147–148
interculture operation, 147
nutrient management (manures and fertilizers), 146
nutritive value, 150–151
origin and distribution, 141–142
plant characteristics, 142–143
production area, 142
seed
inoculation, 145
production, 150

rate, 145
sowing method, 145
treatment, 145
soil and preparation, 143–144
sowing time, 145
spacing, 145
toxicities (antinutritional factors), 151
utilization, 151
varieties, 144–145
water management, 147
weed management, 147
yield, 150

N

Nonconventional legumes forage crops, 287
centrosema pubescens, 288
cultivation practices, 289–290
harvest and storage, 291
utilization, 289
clitoria ternatea, 293
cultivation practices, 294–295
harvesting, 295
utilization, 294
yield, 295
Desmodium, 295
cultivation practices, 297–298
harvesting, 298–299
utilization, 296–297
yield, 298–299
lablab or dolichos bean
cultivation practices, 306–307
harvesting, 307
utilization, 305–306
yield, 307
macroptilium atropurpureum, 299
cultivation practices, 300–301
harvesting, 301
utilization, 300
yield, 301
mucuna pruriens, 301
cover crop, 302–303
cultivation practices, 303–304
harvesting, 304–305
nutritional attributes, 303
planting, 304
soil improver, 302–303
utilization, 302
yields, 304–305
pueraria phaseoloides
cultivation practices, 292
harvest and storage, 292–293
utilization, 291
Nongraminaceous forages, disease management
Amaranthus diseases
anthracnose, 335
damping-off, 336
wet rot, 336
carrot diseases, 329
alternaria leaf blight, 330
bacterial leaf blight, 330–331
black rot, 331
cavity spot, 331
cercospora leaf blight, 332
cottony rot (sclerotinia rot), 332
damping-off, 332–333
downy mildew, 333
powdery mildew, 333
root knot nematodes, 333–334
soft rot, 334
Turnip mosaic virus, 334
rapeseed and mustard diseases
alternaria blight (alternaria brassicae), 324–325
downy mildew (peronospora brassicae), 326
white blister (albugo candida), 326
sunflower diseases
alternaria blight and its management, 322–323
downy mildew and management, 323
powdery mildew and management, 324
rust and management, 324
verticillium wilt and management, 323
turnip diseases, 326
alternaria leaf spot, 327
anthracnose, 327
black root, 327–328
black rot, 328
cercospora leaf spot (frogeye leaf spot), 328
Nonleguminous, disease management

Amaranthus diseases
anthracnose, 335
damping-off, 336
wet rot, 336
carrot diseases, 329
alternaria leaf blight, 330
bacterial leaf blight, 330–331
black rot, 331
cavity spot, 331
cercospora leaf blight, 332
cottony rot (sclerotinia rot), 332
damping-off, 332–333
downy mildew, 333
powdery mildew, 333
root knot nematodes, 333–334
soft rot, 334
Turnip mosaic virus, 334
rapeseed and mustard diseases
alternaria blight (alternaria brassicae), 324–325
downy mildew (peronospora brassicae), 326
white blister (albugo candida), 326
sunflower diseases
alternaria blight and its management, 322–323
downy mildew and management, 323
powdery mildew and management, 324
rust and management, 324
verticillium wilt and management, 323
turnip diseases, 326
alternaria leaf spot, 327
anthracnose, 327
black root, 327–328
black rot, 328
cercospora leaf spot (frogeye leaf spot), 328
clubroot, 328–329
downy mildew, 329

P

Pangola grass (digit/woolly finger grass)
binomial name, 39–40
botanical classification, 39–40
climatic requirements, 41
common name, 40
crop mixtures, 43
description, 41
drainage, 44
fertilizers application, 43–44
geographical distribution, 41
green fodder yield, 44–45
harvesting, 44
intercultural operation, 44
irrigation, 44
land preparation, 42
manuring, 43–44
nutritive value, 45
plant characteristics, 42
soil requirements, 41–42
sowing, 42
spacing, 43
synonyms, 40
utilization, 45
weed management, 44
Para grass (buffalo grass)
botanical classification, 13
botanical name, 14
climatic requirements, 15–16
common name, 14
crop mixture, 17
green fodder yield, 19
harvesting, 18–19
insect-pest and diseases, 18
land preparation, 16
nutrient management, 17
nutritive value, 19
origin and distribution, 15
plant characteristics, 15
seed rate, 16–17
soil requirements, 16
sowing method, 16–17
synonymous, 14
use, 19
varieties, 18
water management, 17–18
weed management, 18

R

Rhodes grass (Abyssinian Rhodes grass)
botanical classification, 65–66
botanical name, 66

climatic requirements, 67–68
common name, 66
crop mixture, 71–72
disease management, 72
green fodder yield, 72
harvesting, 72
insect-pest, 72
nutrient management (manures and fertilizers), 71
nutritive value, 73
origin and distribution, 67
plant characteristics, 67
seed
inoculation, 70
production, 72
rate, 70
sowing method, 70
spacing, 70
treatment, 70
soil and preparation, 68
sowing/planting time, 69–70
utilization, 73
varieties
callide, 69
finecut, 69
katambora, 69
pioneer, 68
topcut, 69
water management, 71
weed management, 71

S

Seed
inoculation, 25–26, 124, 231
production, 28, 62, 127, 234
rate, 26, 59, 124, 231
sowing method, 26, 59, 124, 231
treatment, 25–26, 124, 231
treatments, 195
Senji (sweet clover), 131
botanical classification, 132
botanical name, 132
climatic requirement, 133
cropping systems, 135–136
disease management, 136–137
harvesting, 137
insect-pest, 136–137
nutrient management (manures and fertilizers), 136
interculture operation, 136
water management, 136
weed management, 136
nutritive value, 137
origin and distribution, 133
plant characteristics, 133
seed
inoculation, 135
production, 137
rate, 135
sowing method, 135
treatment, 135
soil and its preparation, 133–134
sowing time, 134–135
spacing, 135
utilization, 137–138
varieties, 134
yield, 137
Setaria grass (African grass)
botanical classification, 3–4
botanical name, 4
botany of plant, 5
climatic requirements, 5
common name, 4
crop mixtures, 6–7
green fodder yield, 9
harvesting, 9
intercultural operation, 8
land preparation, 6
limitations, 10
manures and fertilizers application, 8
nutritive value, 10
origin and distribution, 4–5
palatability, 10
pest and diseases, 8–9
seed
production, 10
seed rate, 6
soil requirements, 5–6
sowing, 6
spacing, 6
toxicities, 9–10
utilizations, 10
varieties

Kazungula, 7
Nandi mark 2, 7
Nandi setaria, 7
Narok setaria, 7
water management, 8
weed management, 8
Soybean (bhatman)
botanical classification, 101
botanical name, 102
classification
Hertz, 104
manchurian, 104
Martin's, 104
climatic requirement, 106
common name, 102
cropping systems
crop mixture, 107–108
crop sequence, 107–108
intercropping, 107–108
description, 103
disease management
aerial blight, 112
alternaria leaf spot, 112
anthracnose, 112
bacterial blight, 113
bacterial pustules, 113
charcoal rot, 111–112
rust, 112–113
soybean mosaic, 113
yellow mosaic, 113
harvesting, 116
insect pest management
gram pod borer, 115–116
green semilooper, 115
griddle beetle, 115
hairy caterpillar, 114
leaf roller, 114–115
stem fly feed, 114
tobacco caterpillar, 115
whitefly, 114
nutrient management, 110
nutritive value, 117–118
origin and distribution, 105
plant characteristics, 105–106
seed
inoculation, 109
production, 116–117
rate, 109
sowing method, 109
treatment, 109
soil and preparation, 106–107
sowing time, 108–109
utilization, 117
varieties, 107
water management, 110
weed management, 111
yield, 116
Subabul (river tamarind)
after care, 246
botanical classification, 241–242
botanical description, 243
botanical name, 242
climatic requirements, 245
common name, 242
nutritive value, 247
pests and diseases, 246
planting, 246
pretreatment, 245
seed collection, 245
soil requirements, 244–245
spacing, 246
toxicities, 248–249
types of, 243
cunningham, 244
Hawaiian type, 244
PERU, 244
salvador type, 244
varieties, 247
biscuits, 248
FD 1423, 248
subabul co-1 (P), 248
yield and rotation, 247
Sunflower (sujyomukhi), 171
botanical classification, 172
botanical name, 172
climatic requirement, 174
common name, 172
compatibility, 183–184
crop sequence, 177–178
cropping systems, 177
description, 173
distribution, 173–174
harvesting, 180–181
intercropped, 177

nutrients management, 178
nutritive value, 182
origin, 173–174
plant characteristics, 174
plant protection
pests and diseases, 179–180
seed
production, 181
rate, 176–177
sowing method, 176–177
treatment, 176
soil and preparation, 175
sowing time, 175–176
special features, 183
utilization, 182–183
varieties, 175
water management, 178–179
weed management, 179
yield, 181

T

Turnip (salgam)
botanical classification, 207
botanical name, 208
climatic requirement, 209
common name, 208
crop mixture, 211
disease management, 212
harvesting, 212
insect-pest management, 212
nutrient management, 211
nutritive value, 213
origin and distribution, 208–209
plant characteristics, 209
seed
bed, 211
inoculation, 210
production, 213
rate, 210
sowing method, 210
treatment, 210
soil and preparation, 209–210
sowing time, 210
spacing, 211
toxicity, 213
utilization, 213
varieties, 210
water management, 212
weed management, 212
yield, 212

U

Urd bean (black gram)
botanical classification, 155–156
botanical name, 156
climatic requirement, 158–159
common name, 156
cropping systems, 161
diseases management, 163–164
harvesting, 164
insect-pests management, 162–163
interculture operation, 162
nutrient management (manures and fertilizers), 162
nutritive value, 165–166
origin and distribution, 157
plant characteristics, 158
production area, 157–158
seed
inoculation, 160
production, 165
rate, 160
sowing method, 160
treatment, 160
soil and preparation, 159
sowing time, 160
spacing, 160–161
toxicities (antinutritional factors), 166
utilization, 166
varieties, 159–160
water management, 162
weed management, 162
yield, 165

V

Vegetative propagation
air layering, 266
budding, 266
cutting, 265–266
micropropagation, 266–267
root suckers, 265
rootstock, 267

Printed and bound by CPI Group (UK) Ltd, Croydon, CR0 4YY
23/10/2024
01777703-0004